Jürgen Audretsch (Editor)
Entangled World

The Fascination of Quantum Information and Computation

Related Titles

Stolze, J., Suter, D.

Quantum Computing

A Short Course from Theory to Experiment

255 pages with 82 figures
2004
Softcover
ISBN 3-527-40438-4

Leuchs, G., Beth, T. (eds.)

Quantum Information Processing

347 pages with 112 figures and 1 tables
2003
Hardcover
ISBN 3-527-40371-X

Entangled World

The Fascination of Quantum Information
and Computation

Jürgen Audretsch (Editor)

WILEY-
VCH

WILEY-VCH Verlag GmbH & Co. KGaA

Prof. Dr. Jürgen Audretsch
University of Konstanz
Department of Physics
Konstanz, Germany
Juergen.Audretsch@Uni-
Konstanz.de

Original title:
Verschränkte Welt —
Faszination der Quanten
© 2002 WILEY-VCH Verlag
GmbH, Weinheim

Translation:
Rudolf Ehrlich, Germany

Library of Congress Card Nº:
applied for

**British Library
Cataloging-in-Publication Data:**
A catalogue record for this book is
available from the British Library.

**Bibliographic information
published by
Die Deutsche Bibliothek**
Die Deutsche Bibliothek lists this
publication in the Deutsche
Nationalbibliografie; detailed
bibliographic data is available in the
Internet at http://dnb.ddb.de

© 2006 WILEY-VCH Verlag GmbH &
Co. KGaA, Weinheim

Typesetting   Hilmar Schlegel, Berlin
Printing and Binding   Ebner &
Spiegel GmbH, Ulm
Cover Design   Himmelfarb,
Eppelheim,
www.himmelfarb.de

Printed in the Federal Republic of
Germany

Printed on acid-free paper

ISBN-13: 978-3-527-40470-4
ISBN-10: 3-527-40470-8

The Authors

Prof. Dr. Jürgen Audretsch
Universität Konstanz
Fachbereich Physik
Universitätsstrae 10
D-78457 Konstanz
juergen.audretsch@uni-konstanz.de

Prof. Dr. Rainer Blatt
Universität Innsbruck
Institut für Experimentalphysik
Technikerstrae 25
A-6020 Innsbruck
Rainer.Blatt@uibk.ac.at

Prof. Dr. Michael Esfeld
Department of Philosophy
University of Lausanne
CH-1015 Lausanne
michael-andreas.esfeld@unil.ch

Prof. Dr. Carsten Held
Philosophie
Universität Erfurt
Postfach 90 02 21
D-99105 Erfurt
carsten.held@uni-erfurt.de

Dr. Erich Joos
Rosenweg 2
D-22869 Schenefeld
ej@erichjoos.de

Dipl. Phys. Robert Löw
Universität Stuttgart
5. Physikalisches Institut
Pfaffenwaldring 57
D-70550 Stuttgart
r.loew@physik.uni-stuttgart.de

Prof. Dr. Tilman Pfau
Universität Stuttgart
5. Physikalisches Institut
Pfaffenwaldring 57
D-70550 Stuttgart
t.pfau@physik.uni-stuttgart.de

Prof. Dr. Gerhard Rempe
MPI für Quantenoptik
Hans-Kopfermann-Str. 1
D-85748 Garching
gerhard.rempe@mpq.mpg.de

Prof. Dr. Harald Weinfurter
Universität München
Sektion Physik
Schellingstr. 4
D-80799 München
harald.weinfurter@physik.uni-muenchen.de

Prof. Dr. Reinhard F. Werner
TU Braunschweig
Institut für Mathematische Physik
Mendelssohnstr. 3
D-38106 Braunschweig
r.werner@tu-bs.de

Contents

Preface

The first time that one of the basic ideas of quantum theory was expressed in public is known exactly to the day. On the 14th of December 1900, Max Planck was introducing the concept of a discrete "energy element" or "energy quantum" into physics in a lecture about the radiation of a black body at the Physical Society in Berlin. In this lecture, he postulated the proportionality of the energy E and the frequency ν of light, which is mediated by a new universal constant, the so-called "Planck's constant" (in German "Wirkungsquantum", which means "action quantum") h:

$$E = h\nu$$

Even before Albert Einstein established the equivalence of mass and energy

$$E = mc^2$$

in the theory of special relativity in 1905, one of the two relations that can be regarded as the symbols of the 20th century physics was already formulated. Later, $E = mc^2$ was conquering posters and color postcards as an icon, while Planck's equation has never reached such a widespread popularity. This is basically surprising, because the quantum theory may possibly have enforced a more profound change of view regarding our idea about the nature of physical reality than the theory of special relativity. The fact that the quantum-physical concepts and notions, which were developed in this process, have such few similarities with those of everyday physics — the figure of the non-classical skier demonstrates this — is the reason for the difference in the popularity of these two equations.

Certainly the year 1900 is accepted as the year of birth of quantum theory, but only in 1925 to 1927 did W. Heisenberg, M. Born, E. Schrödinger, P. A. M. Dirac, W. Pauli, and J. von Neumann succeed in formulating and interpreting the theory in a convincing way.

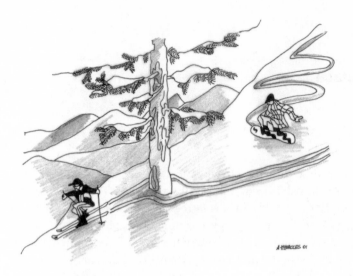

Which path? The nonclassical skier.
(Drawing: A.-M. Herckes)

Today, after three quarters of a century, quantum physics has established itself in our everyday life by taking a detour over technology, however, without being accepted as everyday physics. The reason that we have such little awareness about to what extent we are surrounded by technical applications of quantum physics is due to the fact that we have locked up quantum physics in "black boxes", which we can handle and operate without knowing what is going on inside. We are able to mount a fluorescent tube in a lamp holder and by pressing the light button, let the electric current flow from an atomic power station. Whatever is going on inside the fluorescent tube or in the atomic power station, we (in most cases) do not know. With our actions, we remain within the scope of everyday physics, which is classical physics. Here we are well acquainted with the phenomena, which we believe we understand. Quantum physics — the basis of the modern technology from lasers to microchips — on the other hand, remains an unknown land for many, about which strange and apparently paradoxical things are reported from time to time. And yet, it is not so difficult to obtain some insight into the quantum world with its fascinating phenomena and astonishing theoretical ideas. This book aims to give some stimulation in this direction.

However, since the beginnings of quantum physics were such a long time ago, has this field not yet been investigated, discussed and clarified completely in all details? This question reminds us of a well-known anecdote from the history of physics. When the young Max Planck asked the physicist von Jolly from Munich whether it makes any sense to study physics, he was advised against it: all the essential problems in physics are already explored and at best there might be a gap to be closed here and there. We have already seen that Max Planck did well by not following this advice. Also, in view of quantum physics, the long-time predominating impression that the basics were already completely clarified and the front line of physical research would long be somewhere else was wrong. Not the least because one has understood how to isolate and to manipulate single quantum objects like atoms, ions, molecules or photons, that the old gedanken experiments of quantum theory have become experimentally realizable and have led to completely new insights. The old fundamental questions, which remained open for a long time — for example about the physics of the measuring process — could at last be investigated experimentally and technological developments with far-reaching consequences are indicated, which nobody would have anticipated in this form.

Today there is an euphoric atmosphere of departure. Almost no week goes by without quantum-physical topics on the science pages of the large daily newspapers: quantum computers, quantum information processing, quantum teleportation, quantum cryptography, quantum cats and so on. These "quantum articles" are joined by popular descriptions on the pages on natural philosophy, in which arguments of the type "quantum physics has shown us that anyway everything is connected with each other and therefore ..." are used. All these reports have in common that a key concept of quantum theory is at the center of attention: entanglement. Quantum physics has transformed into an "entangled world". This book is meant to give an insight into this world. The volume is based on lectures about "physics and philosophy of correlated quantum systems" that were held at the University of Konstanz in the winter semester 2000/2001.[1]

---

1) The chapter 9 has been added which was especially written for the English edition.

Similar to a previous volume[2], the authors reporting in this book about the research have actively participated in it. The latest results from the field of the basics of quantum physics, as well as the challenges on the way to technological applications are presented with special emphasis on the last ten years. Beyond that, efforts are made to answer the central philosophic question about the nature of reality. The problem of having a thematic overlap between some of the articles has been accepted on purpose, to produce better intelligibility. Corresponding cross-references were added.

One should not miss the fact that a clear definition and use of terms, as well as an appropriate formulation of the questions and results are mandatory in order to understand the quantum world. This does not mean that the reader should have studied physics before reading this book. Surprisingly, a good idea about the basic structures of quantum theory can be gained with only a little previous knowledge in mathematics. Elementary concepts like vector and scalar product are nevertheless needed, when trying to follow the line of argument or to think ahead without getting tangled up in alleged contradictions and pseudoproblems. The ideas and conceptions of quantum theory are unfamiliar to us, because quantum theory does not describe and explain phenomena of the well-known everyday physics, which are directly accesible to us. This is done by classical physics. Quantum theory compels us to give up our usual way of thinking. Also for this reason, it has such great attraction for us.

The many aspects of entanglement are topics that are now step by step entering the curricula for the upper courses of secondary schools and the ground courses at universities. Also there, this book may be useful. This book is intended for students, natural scientists, engineers, philosophers, and in particular for pupils and teachers and not least for the interested "layperson". Beyond that, it is meant to contribute to the interdisciplinary dialog between physics and philosophy by stimulating an examination of the reality of the entangled world of quanta.

Konstanz, Autumn 2001   *Jürgen Audretsch*

---

2) Audretsch and Mainzer (1990/1996).

1

View into the quantum world I: fundamental phenomena and concepts

Jürgen Audretsch

1.1 Introduction

With this chapter and the following one, which has the title "Entanglement and its consequences", an introduction into the world of quantum physics and its description by quantum theory will be given in a self-contained way. At the same time, the understanding of Chapters 3 to 10 should be made easier with this introduction. For didactic reasons, we will always return to two basic experiments — which used to be just gedankenexperiments for a long time — the transition of quantum objects through a single slit and through a double slit. This limitation should not be misunderstood. Apart from some exceptions, all experiments that are performed in experimental physics nowadays are based in one way or the other on quantum-physical phenomena. To be able to read the structures of quantum theory particularly clearly, it is, however, recommended to start from simple experiments.

The demands are growing from section to section in both chapters.[1] In Sections 1.2 to 1.5, fundamental experiences with quantum objects are described — based on the example of the slit and double-slit experiments. The approaches for the theoretical representation of quantum objects, which is the topic of Section 1.6, are also introduced. Thereby, however, only those elements of quantum theory that are actually required for the following sections are formulated. The quantum Zeno

---

[1] Who wants to continue the ascent can find a more detailed and precise presentation of the whole subject for instance in the university test book Audretsch (2005).

Entangled World, Jürgen Audretsch
Copyright © 2005 WILEY-VCH Verlag GmbH & Co. KGaA, Weinheim
ISBN: 3-527-40470-8

effect described in Section 1.7 is a first application. This effect can be understood without any reference to mathematical relations. After introducing photons as the quantum objects of light in Section 1.8, the theoretical formalism can be illustrated for the first time with the example of interaction-free quantum measurements in Section 1.9. We will discuss the resulting picture about the quantum world in the concluding Section 1.10. The question about the structure of the reality in quantum physics will be at the center of this discussion.[2]

Anyone trying to understand quantum physics will quite soon be confronted with the question what the expression "to understand" really means in this context. Quantum physics is exactly not classical physics. Trying to understand quantum physics by transferring classical pictures is therefore useless. Classical physics can only be used by way of comparison in order to make particularly clear what is different in quantum physics. We will return to this problem several times again. The reader might become encouraged when learning that such difficulties of understanding are not uncommon at all. The famous physicist Lord Kelvin (1824–1907), who made important contributions to the theory of heat, wrote in 1884:

> "I am never content until I have constructed a mechanical model of the object that I am studying. If I succeed in making one, I understand; otherwise I do not. Hence I cannot grasp the electromagnetic theory of light."
> (Mason (1953))

Obviously Lord Kelvin already had to come to terms with the fact that electrodynamics could not be reduced to classical mechanics any longer.

Just one more word regarding the mathematical requirements: up to Section 1.5 we can do almost without mathematics. Strictly speaking, complex functions $\Psi(\underline{r}, t)$ and the formation of their absolute value $|\Psi(\underline{r}, t)|$ are necessary for the description to be correct. For a first understanding of the structure of quantum theory, it is totally sufficient to think of $\Psi(\underline{r}, t)$ as a real function. The same holds for Section 1.6. The visualization of the vectors in Fig. 1.8 is also carried out in the real vector space. Only in one place — which is in the

---

2) Regarding the literature quotations, no completeness is claimed. Reprints of the most important works can be found in Wheeler and Zurek (1983) and Macchiavello et al. (2000).

second half of Section 1.9 — is the imaginary number i actually introduced into a calculation, in order to represent the phase shift. Also in this case though, the physical effect of an interaction-free measurement is deduced first of all without physical equations. For further calculations however, complex vector spaces are required. From vector analysis, the vector addition and the scalar product (inner product) of two vectors are used starting with Section 1.6. A new terminology — compared to classical mechanics and electrodynamics for example — is going to be used for these calculations out of practical and historical reasons.

1.2 Diffraction at a single slit

The elementary building blocks of matter, like electrons, neutrons, protons etc. are called elementary particles. Therefore, they seem to be particles. At the same time one can frequently hear that these particles are supposed to have a wave nature. Can particles be waves at the same time? This is simply hard to imagine. Or are these elementary particles in certain situations behaving like particles and in others like waves? We will see that this conjecture is not totally wrong. Although it is expressed in a way that could be misunderstood and it will be our task to develop stepwise the precise ideas and formulations that let us describe experiments with elementary particles, including photons and also atoms and molecules. The theory that achieves this is the *quantum theory*. When the theory is limited to the description of objects with mass it is also called *quantum mechanics*. While studying the problem in a more systematic way, how can one actually get to the idea that the wave concept plays any role for the description of the physics of objects with small masses?

For a long time, the physics of massive objects has been described with great success by mechanics, which we more precisely call *classical mechanics*. Light is a phenomenon based on electromagnetic fields. We are going to start from these two *classical* theories and, in a first step, comment on two well-known experiments.

At first we give an account of an everyday experience, which can be described using classical mechanics. Tennis balls are flying in a perpendicular direction towards a wall in which we have a window opening. The tennis balls that come through hit the opposite wall. We mark the points where they hit this wall and determine the relative frequency $P_{cl}(x)$ with which these positions x are hit. In the case of a homogeneous current of tennis balls, one gets the curve in Fig. 1.1: we find hit points only directly opposite to the window opening. The same experiment can be repeated with beams of light. Again, this results in a regular distribution of brightness opposite to the window opening. The linear light rays at the edge define the shadow regions. This phenomenon can be explained by *geometrical optics.*

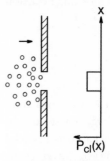

Fig. 1.1 Tennis balls flying through a slit are registered with the relative frequency $P_{cl}(x)$ on an opposite screen.

The propagation of light is in fact a wave process. The wave properties of light will show up when the dimension of the opening the light comes through is of the order of magnitude of the wavelength. Then the effects of *diffraction optics* replace geometrical optics. We will discuss this while looking at a single slit, which is lit by a plane light wave (Fig. 1.2). An intensity distribution $P_1(x)$ results now on a photoplate (screen) behind the slit The distribution shows a maximum directly opposite the center of the slit, and side maxima that are separated from each other by minima with vanishing intensities beyond the shadow limits. This is a diffraction image, which results from *interference.* Spherical waves caused by the plane wave that strikes the slit are sent off from different positions in the slit and they overlap behind the slit. When a wave maximum meets a wave maximum with the overlap at a given position, a higher wave maximum is created, and in the case of two wave minima a lower wave minimum correspondingly. When a wave maximum meets a wave minimum they cancel each other out. The wave picture behind the slit is therefore the result of an addition. This is also called the *superposition* of the elementary waves in this context. The intensity distribution

$P_1(x)$ can be recorded for example with the blackening of a photo-plate. The plate, though, senses wave maxima and wave minima in the same way, therefore only the square of the wave amplitude gives a measure for the blackening. In short, one might say that the interference pattern is obtained following the rule "add first, then square". $P_1(x)$ in Fig. 1.2 shows the resulting relative frequency of blackening points.

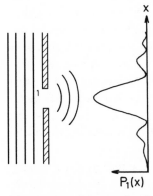

Fig. 1.2 A plane light wave (symbolic wave trains) hits a slit 1. The diffraction pattern $P_1(x)$ gives the intensity distribution, which is registered on an opposite screen (photoplate).

The appearance of a diffraction pattern is a direct indicator of the occurrence of interference. The result "light + light = darkness" cannot be produced with particles. Therefore, it must be mathematically a wave phenomenon that underlies this. Still, before discussing this we have to point out that geometrical optics and diffraction optics are two theories of light independent of each other, each one of them to be applied in special physical situations. The *electrodynamics*, that is the Maxwell Theory of electromagnetic fields from 1873, which also describes all light phenomena in a unified way, is a diffraction theory from its structure. The theory of geometrical optics is included as a limiting case. Electrodynamics, therefore, is the more general and comprehensive theory, but in special physical situations one can speak about light propagation along light beams to a good approximation.

Now we will return to our initial question, whether diffraction phenomena and thus interference exist also for molecules, atoms, elementary particles etc. In analogy to optics, we might expect the following relation between theories

| geometrical optics | ← | diffraction optics |
| classical particle physics | ← | ??? |

The arrows indicate the "transition in the limiting case". The following question arises now: is there any general theory for all material objects from elementary particles to stars, for which the superposition and the interference are central concepts and that, in certain physical situations, leads back to classical mechanics as a limit with its well defined particle paths? This theory would then be a general theory for the whole mechanics, which would also predict and describe a whole host of new phenomena. These are not necessarily diffraction effects in the strict sense because interference causes more than just diffraction. With *quantum mechanics,* such a theory has indeed been available for about 75 years. We name the objects described there *quantum objects* in order to avoid the misleading expression "particle".

Still, we should be cautious about using the analogy to electromagnetic phenomena. We will essentially just read from the similarities described above that in both cases interference and with that superposition play a central role in the mathematical description. We should expect that beyond this similarity quantum mechanics and electrodynamics are clearly different from each other both conceptually and formally. In particular, one theory will not be reducible to the other. It is rather the case that a quantum structure is also hidden behind electrodynamics. The corresponding quantum objects are the photons. Their existence and their quantum behavior are extraordinarily well confirmed with a whole host of experiments in high-energy physics and in quantum optics. We will return to this later. Since more than just objects with a mass are described, it is correct to speak about *quantum* theory rather than quantum mechanics.

However, we should first ask ourselves: do interference phenomena occur at all in connection with material objects?

1.3 Atom optics

During the first years of quantum mechanics it was just a gedanken experiment. In the meantime, it has been possible for many years to demonstrate diffraction phenomena for electrons, neutrons, atoms and molecules directly in experiments.[3] As for light, the diffracting arrangements like single slit, double slit or grating have to be suitably dimensioned. We are going to outline the scheme of such an experiment again for the example of a single slit. It is remarkable that almost all theoretical and conceptual elements leading towards the basic assumptions of quantum theory can be studied with this very clearly laid out experimental setup together with the measurement results that can be described in a very simple way. Therefore, it is justified that the diffraction of material objects at a slit or at a double slit have very often been chosen as the starting point for a description of quantum mechanics.[4]

We now take a look at atoms, all prepared with equal momentum, hitting a wall with a slit from a perpendicular direction. A screen on which the impact of the atoms can be registered, is located behind the slit and parallel to the wall. The details of this are irrelevant. Let us assume that the impact is documented by a blackening. When the incident current of atoms is not too dense, blackening spots of point shape are registered, which appear one after the other without regularity and thus randomly distributed over the screen (Fig. 1.3). It is possible to dilute the beam so much that there is always only one atom inside the setup and thus its registration on the screen takes place before the next atom is flying towards the slit.

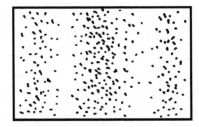

Fig. 1.3 Impact points of quantum objects (e. g. atoms) that passed through a slit show a blackening image in accordance with the diffraction at a slit.

3) A review about the experiments with atoms can be found in Berman (1997). For neutrons and electrons see Rauch and Werner (2000) or Tonomura (1998), respectively.
4) An example of this is Feynman et al. (1965).

Already at this point we can draw the first conclusions. Since single impacts are observed, we will be able to speak meaningfully about *single* quantum objects, in this case single atoms, which were inside the installation. A *position measurement* of a single quantum object is carried out on the screen due to the fact that we can exactly determine the position of the point-shaped impact. After collecting the impacts of many quantum objects, the resulting blackening picture looks not at all irregular. Amazingly one finds for the intensity of the blackening or the number of impacts per area element exactly the same distribution as in the case of light diffraction at a slit (Fig. 1.2). We have therefore obtained a diffraction image for quantum objects, too.

It is notable and actually even more surprising, that exactly the same intensity distribution can be realized experimentally in three very different ways:

1. For a specific arrangement of slit and screen, a dense current of quantum objects can be sent through a slit, so that many quantum objects hit the screen at the same time.

2. The current can be thinned out, so that there are always only single quantum objects one after the other inside the installation and hitting the screen.

3. Finally a huge number of the same experimental arrangements consisting of slit and screen can be built, and just exactly one quantum object is sent through each of these single arrangements and all the single hits from the many plates are marked in one single graph.

In all three cases we obtain the blackening distribution described above. Still, there has always been some chance at work. When, for example, the second procedure is repeated, the places hit by the first, second, etc., object are totally different each time. Nevertheless, the resulting overall picture after many impacts is again the same. We call this an *ensemble experience*.

We summarize once again: in all three cases it is guaranteed that the quantum objects cannot interact with each other. A single object, therefore, has absolutely no information about the places where other quantum objects have already hit the screen. Also, there is no prediction about the place where a single object is going to hit. The single impact takes place in an *undetermined*, which means random way. When the experiment is repeated with many quantum objects, still the same diffraction image emerges again and again. The formation of this overall picture is therefore a *deterministic* process. The overall picture can be precisely predicted. All these observations are of course *objective*. The experiments can in principle be repeated by anyone anywhere in the world, with the same result.

How can we describe these phenomena theoretically? The missing prediction in the single case and the definite prognosis for the combination of many results are well known from throwing dice. There, one can also throw the same dice several times or throw only once with several dice. With good approximation, $1/6$ of the cases will give, for example, the number 1. The *relative frequency* of the number 1 is $1/6$. When we want to make a prediction for the result of a single throw we say that the *probability* of getting for example the number 1 is $1/6$. We now transfer this form of a description cautiously to our quantum objects. While doing so, we must not conclude that there is anything behind the diffraction process of quantum objects that would be physically similar to the process of throwing dice in classical mechanics. We have no reasons for this and in fact we are going to show that this is really not the case.

For the mathematical description we will combine now the elements of determination and indetermination. We introduce a function $\Psi(\underline{r})$ of the position $\underline{r}$,[5] which describes the wave situation behind the slit. Similar to the electromagnetic field, the diffraction at the slit is expressed by the particular spatial behavior of this function. The square of the absolute value of the function, $|\Psi(\underline{r})|^2$, is interpreted physically as the impact probability $P(\underline{r})$ at a position in a small volume element dV around a position $\underline{r}$. This may be regarded as the prediction of the relative frequencies of the results of a position measurement. In other physical situations, Ψ and P will also depend on the time t. The function $\Psi(\underline{r})$ is called the *state function* or *Schrödinger function*. The screen can also be placed at different positions behind the slit. In any case, $|\Psi(\underline{r})|^2$ with the corresponding position vector $\underline{r}$ will give the correct impact probability on the screen. The state function describes the physical situation behind the slit but before the measurement. The function is determined by the

---

5) A complex time-dependent factor with an absolute value of one is suppressed.

width and the position of the slit. Different apertures will lead to different functions. One can also say that the function $\Psi(\underline{r})$ describes a specific *quantum state*.

In this description, we are strictly limiting ourselves to the prediction of position measurements. The state function serves only this purpose. Behind the slit there is no vibrating physical quantum substance or a "wave-pudding" of quantum objects imagined as being smeared out. The name "matter wave" for $\Psi(\underline{r}, t)$ is in this sense unfortunate. There are no classical particles with position and momentum, because in this case one should expect the result of Fig. 1.1. Instead, we speak about one or many quantum objects with an assigned quantum state, which is described by the function $\Psi(\underline{r}, t)$; $\Psi(\underline{r}, t)$ may have mathematically the form of a wave in certain situations.

After discussing the ensemble experience and formulating the elements of quantum theory, we shall propose, on this basis, a prognosis for a modified experimental arrangement. We let atoms fly, but instead of using one slit, we put two similar, aligned parallel and suitably dimensioned slits in their way. Our conjecture is that the analogy to the diffraction image of light waves passing through a double slit[6] will become evident in the resulting frequencies of impacts on different places of the screen. The superposition principle applies to light waves; this means that the single light waves, coming from slit 1 and slit 2, interfere behind the double slit. They add up as discussed above. When the square of the resulting wave field is determined, for example at the positions x on the screen, the normalized intensity distribution of the field $P(x)$ from Fig. 1.4 is obtained and thus the relative frequency of blackening points on the photoplate.

When atoms instead of light are incident on a double slit, a totally analogous image for the frequency of impacts of atoms on positions of the screen emerges. Again all three ensemble experiences are valid. This result is described in the same way by superposition, but in this case the state functions $\Psi_1(\underline{r})$ and $\Psi_2(\underline{r})$ are to be added. $\Psi_1(\underline{r})$ is the state function present when slit 2 is closed. This is the state function behind a single slit as discussed above. $\Psi_2(\underline{r})$ is the corresponding state function when slit 1 is closed. Therefore, we have superimposed the quantum state "through slit 1" and the quantum state "through

6) A modern double-slit experiment for atoms is described in this book by
G. Rempe in Section 5.1.

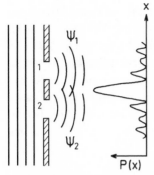

Fig. 1.4 The intensity distribution $P(x)$ of quantum objects diffracted at a double slit shows a maximum opposite to the bridge. The symbolic wave trains of the state function before and after the slit are drawn. Ψ_1 (Ψ_2) is the state function that is present when slit 2 (1) is closed. Ψ_1 and Ψ_2 are superimposing.

slit 2".[7]

$$\Psi(\underline{r}) = \frac{1}{\sqrt{2}} \left(\Psi_1(\underline{r}) + \Psi_2(\underline{r}) \right) \tag{1.1}$$

$$|\Psi(\underline{r})|^2 = \frac{1}{2} |\Psi_1(\underline{r})|^2 + \frac{1}{2} |\Psi_2(\underline{r})|^2 + \frac{1}{2} \left(\Psi_1^*(\underline{r})\Psi_2(\underline{r}) + \Psi_1(\underline{r})\Psi_2^*(\underline{r}) \right) \tag{1.2}$$

The star indicates the complex conjugate function.[8]

Atom optics with a double slit shows in a particularly drastic way that quantum objects do not behave like small classical particles with a mass. The frequency of impacts is not highest behind the two slit apertures but behind the bar in between them. Again we can make the ensemble experiences. The frequency distribution of Fig. 1.4 can also be found in the case where there has been only one quantum object in the setup at a time. Equation (1.2) shows that this curve is not obtained by just adding the shifted curves for the frequency distribution of Fig. 1.2:

$$P(x) \neq \frac{1}{2} \left(P_1(x) + P_2(x) \right) \tag{1.3}$$

$$P(x) = |\Psi(x)|^2, \ P_1(x) = |\Psi_1(x)|^2, \ P_2(x) = |\Psi_2(x)|^2$$

7) The factor $1/\sqrt{2}$ is necessary because all state functions are normalized due to the probability interpretation. We will return to this in Section 1.6.
8) Note the comment about complex functions in the introduction.

Therefore, one cannot say that one half of the quantum objects went through slit 1 and the other half through slit 2. The last two terms in Eq. (1.2), which are not present on the right side of Eq. (1.3) anylonger, are responsible for the formation of the diffraction pattern of Fig. 1.4. Even the single quantum object behaves fundamentally differently at a double slit as compared to a single slit. Has the single quantum object passed through both slits? We will have to study later in detail if such a question that results from a particle picture makes any sense at all.

1.4 The quantum domain

We have spoken previously about the possibility to find a suitably dimensioned slit for the diffraction of atoms. How do we have to proceed? Since we are dealing with a wave phenomenon, this question is about the determination of the wavelength.[9] Let us take a look at the plane waves, which run towards the double slit. A momentum $p = m\underline{v}$ can be attributed to the corresponding quantum object, which has a direction that agrees with the propagation direction of the plane wave. The absolute value of the momentum results from the preparation procedure. The quantum objects run, for example, through an acceleration voltage by which they gain a certain kinetic energy that can be used to calculate the magnitude of the momentum in the usual way. Since the details of the diffraction pattern depend on the wavelength, a relation between the absolute value of the momentum p and the wavelength λ can be read from an experiment. One finds the *De-Broglie relation*:

$$p = h/\lambda \qquad (1.4)$$

This equation can be confirmed for other sorts of quantum objects and in other diffraction experiments. h is *Planck's constant* (or *Planck's "Wirkungsquantum"* in German)

$$h = 6.626 \times 10^{-34}\,\mathrm{kg\,m^2/s^2} \qquad (1.5)$$

[9] It is also possible to enter the quantum world by slowing down and thus cooling heavy particles like atoms. The current research field of the formation and manipulation of Bose–Einstein condensates is introduced in Section 4.3.

$(\text{kg m}^2/\text{s}^2 = \text{Watt s}^2)$. Note that the value of h in the units of the quantities of everyday physics is extraordinarily small.

From the discussions so far we can also read in which way a quantum experiment typically proceeds. It starts at time t_0 with the *preparation* of a very particular quantum state $\Psi(\underline{r}, t_0)$. In our case, this state has mathematically the form of a plane wave with a definite wavelength and a definite propagation direction perpendicular to the slit. This quantum state is then subjected to an alteration by some influence from outside or an interaction. A charged particle for example can be exposed to a position- and time-dependent electrical potential. This is a process that can last for a certain time. The resulting continuous temporal alteration of the state can be calculated by means of quantum theory and it leads to the time-dependent function $\Psi(\underline{r}, t)$. This *dynamical evolution* of the state function as a function of the time t starting from an initial state is well defined and thus *deterministic*. In our case, this is the transition to the state $\Psi(\underline{r})$ of Eq. (1.1). Finally, the *measurement* is carried out. In our case, this is the position measurement on the screen. The state function at the moment of the measurement fixes the relative frequency of the different measurement results. When the experiment is always carried out just for one single quantum object in the setup at a time, it has to be repeated very often and the quantum objects must always be prepared in the same way.

We return now to the problems that we encountered in Section 1.2. The experiments described above have convincingly shown that massive objects exist, with a behavior that can fundamentally deviate from the behavior that we are used to in classical mechanics. Therefore, a *quantum domain* exists in nature. How wide is its range of application and, correspondingly, the range of validity of quantum mechanics extended? We have seen for the diffraction of light at a slit that the beams of the geometrical optics can be used for the projection of the slit, when the wavelength is very small compared to the width of the slit.[10] The pattern with the shadows of Fig. 1.1 emerges then from the diffraction image of Fig. 1.2. Using the De-Broglie relation, a wavelength can also be assigned to the quantum objects, and again one finds the same effect: when the wavelength is very small compared to the slit aperture, a fraction of the plane wave passes through

---

10) Also macroscopic quantum effects exist: superconductivity, superfluidity, Josephson effect, etc.

the slit aperture almost without any modification and the quantum objects cast a "shadow". With that, we are back again to the tennis ball experiment described above.

A quantum phenomenon is transformed to a classical phenomenon when the circumstances are changed. Does this mean that the domains of application of both theories touch without having any overlap? Or does the domain of application of quantum mechanics include classical mechanics (Fig. 1.5)? An extreme case of the first version was supported by the so-called Copenhagen interpretation of quantum mechanics. The research nowadays is pursuing the second approach, in which universal validity is attributed to the quantum theory. This approach follows the basic idea from our first section that electrodynamics also describes geometrical optics. It leads directly to an important, still unsolved problem. We have seen — and this will get clearer later — that quantum mechanics is very structured differently compared to classical mechanics. For macroscopic objects, no superposition of states exists. Cats, for example, are either dead or alive. Nobody has observed so far the hermaphrodite existence of a superposition of both states. How should one therefore proceed within quantum mechanics in order to describe objects of everyday physics? What must be the case so that an object behaves in the "classical" way?[11] For this approach not quantum physics, but classical physics is the unsolved problem. Where is the borderline between the quantum world and the classical world? Fullerenes with 60 and 70 carbon atoms still show diffraction at a grating (Arndt et al. (1999)).

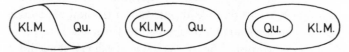

Fig. 1.5 Different possibilities of how the application domains of the classical mechanics (Kl. M.) and the quantum mechanics (Qu.) are related to each other. The middle version is favored in the article.

The third possibility, that quantum mechanics can be traced back completely to deterministic classical mechanics, still needs to be discussed (Fig. 1.5). This is an idea that is very attractive from some philosophical points of view. In the course of the preceding decades were made repeated attempts to base a viable theory on it. Today, it

11) Chapter 8 deals with this question.

can be proved by experiment that this is impossible. We will go into that in more detail in the following chapter. The quantum-mechanical probabilities are not reducible. This is the essential statement. Objective *chance* exists in nature. The situation is in fact *not* the same as for dice, where each throw goes really completely deterministic and only subjectively do we get the impression of chance, and a probabilistic description reproduces our experiences. Albert Einstein in 1949 argued against the probability interpretation of quantum mechanics in a polemic way with the famous saying: "God is not playing dice". He was right in a sense, when the phrase is understood in a different way from how it was meant by Einstein. Then it means that quantum objects are not classical objects like the dice and therefore no probability reducible to deterministic physics exists in this domain.

We would like to mention another consequence. From the fact that quantum mechanics cannot be reduced to classical mechanics follows that the classical concepts fail in the quantum domain. However, we have developed and practiced our intuition facing the everyday physics. When we are not educated in physics, only things that can be described and explained based on classical physics appear to be intuitive and obvious for us. If intuitive knowledge is understood in this way, quantum physics is necessarily *not intuitive or plausible*.[12] Also, colloquial language, which is used to express everyday phenomena, can handle phenomena only from classical mechanics without problems. We will see that quantum physics, in contrast to this, requires a "reduced" language in order to avoid suggesting via the wording propositions that are actually not valid in the quantum domain.

1.5 Quantum measurements

With the position measurement on the screen, we so far only know about a very simple type of quantum measurement. Measurement processes are actually another large field, in which quantum objects behave in a characteristic way that is totally different compared to objects of classical physics. We return to our double slit, which can be used to illustrate this. In Section 1.3, when discussing the double slit, we emphasized the fact that it is just not possible for the interference

---

12) Compare the problem of Lord Kelvin sketched in Section 1.1.

image to emerge from the sum of the probabilities, but instead from the superposition of both state functions Ψ_1 and Ψ_2, which belong to the two single slits 1 and 2. We have seen further that we are not able to decide whether the single quantum object came through slit 1 or through slit 2. Perhaps we have just missed collecting the necessary information by means of a measurement. Now we are going to make up for that.

For this purpose we use electrically charged quantum objects that are able to scatter light and irradiate the space directly behind the screen (Fig. 1.6).[13] We realize then, that for every object a flash occurs either behind slit 1 or behind slit 2 before the impact is recorded on the screen. Thus, we have made a measurement by light scattering to answer the question "Through which slit?" and obtained as a result either "through slit 1" or "through slit 2". A simultaneous flashing behind slit 1 and behind slit 2 never occurs. This would anyway require that the quantum object (for example an elementary particle) could somehow be split in two by the double slit. When taking a look at the impact points of many quantum objects on the screen, well-defined patterns are formed again. When the impacts corresponding to the flashes behind slit 1 are considered separately, exactly the intensity distribution belonging to the state Ψ_1 is obtained (cf. Fig. 1.2). In spite of having slit 2 open, a diffraction image was formed, which totally agrees with the image obtained when slit 2 was closed. In the same way, one gets for the impacts corresponding to the flashes behind slit 2 the diffraction image that belongs to Ψ_2. When all impacts are brought together in one graph, the intensity distributions simply add up. The result does therefore not agree with the interference result for the double slit (Fig. 1.4).

What should be expected from classical physics in contrast? It is characteristic of classical measurements that the measured object is not modified by the measurement. For example, a position measurement using radar has no influence on the state of motion of the object. This is obviously fundamentally different for measurements in the quantum domain. The position measurement with the result "through slit 1" transforms the state Ψ from Eq. (1.1) to the state Ψ_1 behind the screen. The same applies to slit 2. Here the state Ψ_2 is formed. So the state before the measurement is transformed into

---

13) Compare with Section 5.1.

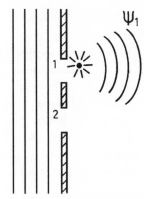

Fig. 1.6 Behind a double slit, the position where a quantum object comes through is measured. For the registration behind slit 1, a diffraction image is obtained on the screen, which corresponds to the state function Ψ_1.

a completely different, new state depending on the measurement result. A well-defined final state belongs to each measurement result that is also named the *eigenstate* of the measured quantity. When we know the measurement result, we also know precisely the state of the quantum object after the measurement. In this sense, a quantum measurement is *preparing* a new state.

When a single object after a first irradiation is irradiated again immediately, it flashes again behind the same slit at the same position like the first time and never behind the other slit. When the experiment is repeated for many objects and only the impacts on the screen belonging to double flashes behind the first slit are regarded, the diffraction image of the single slit of Fig. 1.2 is formed again. The second measurement to answer the question "through which slit?" — which is now made on the state Ψ_1 when the first flash occurred behind the first slit — has not modified this state Ψ_1. When the first measurement has given the result "through slit 1", the directly repeated measurement gives the same result. At least, the fact that an immediately repeated measurement reproduces the first measurement result is common for quantum measurements and classical measurements. One would rather not speak of a measurement, if this had not been the case.

Once again, we return to the single slit and the position measurement on the screen discussed in Section 1.2. We take a look at the quantum state Ψ_1 that belongs to a slit of width Δx (Fig. 1.7). We now carry out position measurements for many objects directly behind the slit. Then all positions behind the slit aperture occur as measuring points with the same frequency. The readings of the coordinate x *scatter* with a width Δx. When we look again at the situation in which there is only one quantum particle in this experimental setup at a time, we can say that the position of the quantum particle before the position measurement is undetermined with the *uncertainty* (square root of the mean square deviation) Δx. For a different experimental arrangement with a slit of the width $\Delta x'$, a state Ψ_1' is present with a positional uncertainty $\Delta x'$ and so on.

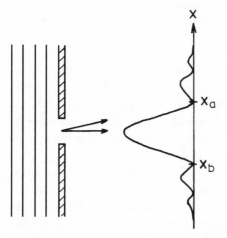

Fig. 1.7 The diffraction at a slit mirrors the uncertainty relation.

We return now to the diffraction image of the single slit (Fig. 1.7) on a distant screen and name the coordinates of the first two minima x_a and x_b. The position of a single impact on the screen can also be considered as an indirect measurement of the momentum direction. The momenta are scattering around the momentum direction of the incoming plane wave (which corresponds to the impact in the maximum) with an uncertainty Δp_x. This Δp_x can be roughly estimated by the distance $x_b - x_a$. It corresponds to an *uncertainty* Δp_x for the momentum.

The evaluation of the two types of experiments with many quantum objects shows

$$\Delta x \cdot \Delta p_x \geq h/4\pi \qquad (1.6)$$

The larger symbol is due to the fact that impact points can be found outside the range between x_a and x_b. The inequality (1.6) is valid for any slit width, and thus also for the state Ψ'_1 with $\Delta x'$ and the accompanying $\Delta p'_x$. The relation (1.6) is named *uncertainty relation*. Uncertainty relations also exist for measurement quantities other than position and momentum.

Sometimes, in this context, the expression indeterminacy is used. This is misleading, because all single position and momentum measurements have resulted in an exact measurement value in each case. We are not dealing with fuzzy measurements done with poor apparatus. The relation (1.6) is rather a statement about the scattering of many measurement results around a mean value using the same preparation of the state. Very often, either the position or the momentum is measured.

When the slit width is reduced, the distance $x_a - x_b$ increases and therefore the diffraction image is stretched. This leads to a limit that is also expressed in Eq. (1.6): $\Delta x \rightarrow 0$, $\Delta p_x \rightarrow \infty$. When a state is such that the result of a position measurement is exactly determined, the result of a momentum measurement is totally undetermined. When the width of the slit is increased, we get just the opposite case. Something else can be directly read from the inequality (1.6): it is not possible that Δx and Δp_x become zero at the same time. No quantum state exists, in which both momentum measurement and position measurement are obtained without scattering.

1.6 A theory for the quantum domain

We are going to summarize the previous experiences and generalize them to form the quantum theory. Such a theory should be founded on just a few basic assumptions and it should be completely formalized from the mathematical point of view. Only then can one clearly realize whether it is logically consistent. Furthermore, a good physical theory is expected to be simple in its mathematical structures and theoretical concepts. For our considerations, we will actually only need the knowledge of complex numbers and some vector algebra in two dimensions.

The quantum state introduced above is a fundamental concept of quantum theory. The use of vectors for the mathematical representation of states was found to be appropriate. The *state vectors* are written in brackets: $|u\rangle$, $|v\rangle$, ... in order to indicate their vector character. The *inner product* (scalar product) of two vectors is written as $\langle u|v\rangle$ and its value is allowed to be a complex number.

As a central physical operation, we have to express the *superposition* of states. This is represented by the vector addition:

$$|w\rangle = a|u\rangle + b|v\rangle \tag{1.7}$$

In general, a and b can be complex numbers. For the formation of the inner product, the dual vector for $|w\rangle$, $\langle w| = a^*\langle u| + b^*\langle v|$, is used. The star indicates the complex conjugate number.

When the passing of the quantum object through slit 1 is measured, the object is afterwards in the state $|\Psi_1\rangle$, and for slit 2 in state $|\Psi_2\rangle$, respectively. When no position measurement is carried out behind the slit, the state $|\Psi\rangle$ behind the slit is a superposition:

$$|\Psi\rangle = c_1|\Psi_1\rangle + c_2|\Psi_2\rangle \tag{1.8}$$

The slit numbers 1 and 2 are related to the possible *measurement results*.

The *probability* that the measurement result 1 is registered results in the theory from the square of the absolute value of an inner product:

$$P_1 = |\langle\Psi_1|\Psi\rangle|^2 \tag{1.9}$$

This applies to measurement value 2 accordingly. When the state $|\Psi\rangle$ from Eq. (1.8) is inserted, one finds that the rule "add first, then square" is expressed, which we learned about in Section 1.2 in the context of interference and diffraction (see also the Eqs. (1.1) and (1.2).

When $|\Psi\rangle$ is present, the probability that an object is found in the state $|\Psi\rangle$ equals one. For this reason it is required that all state vectors, including $|\Psi_1\rangle$, $|\Psi_2\rangle$, etc. are normalized: $\langle\Psi|\Psi\rangle = 1$. When the quantum object is in the state $|\Psi_1\rangle$, the probability to measure the result 2 equals zero. This means that the states $|\Psi_1\rangle$ and $|\Psi_2\rangle$ are orthogonal to each other: $\langle \Psi_2 | \Psi_1 \rangle = 0$. From this, using Eq. (1.8), the relation $|c_1|^2 + |c_2|^2 = 1$ follows, and from Eq. (1.9)

$$P_1 = |c_1|^2, P_2 = |c_2|^2 \tag{1.10}$$

This means in summary that the sum over all probabilities equals one: $P_1 + P_2 = 1$. One of the possible measurement values is found in each measurement.

When both probabilities P_1 and P_2 are the same, the state behind the double slit must have the form:

$$|\Psi\rangle = \frac{1}{\sqrt{2}} \left(|\Psi_1\rangle + |\Psi_2\rangle \right). \tag{1.11}$$

When the passage through one of the slits is made more difficult by means of a filter for example, the probabilities to register the quantum object with a measurement behind one of the slits is no longer the same. This is reflected by the general superposition (1.8).

Except when the scalar product is a complex number, we can use the vector arrows known from vector algebra of the real vector space for the graphical illustration of the states. Figure 1.8 then represents Eq. (1.8). When a filter is installed, we have $\alpha \neq 45°$.

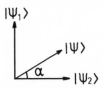

Fig. 1.8 The superposition of the two quantum states $c_1|\Psi_1\rangle$ and $c_2|\Psi_2\rangle$ gives the state $|\Psi\rangle$.

A *position measurement* gives as a measurement result the position vector $\underline{r}$ and transforms the state into the state $|\underline{r}\rangle$. We introduce the state function $\Psi(\underline{r})$, which was already used in Section 1.3 as an abbreviation: $\Psi(\underline{r}) = \langle \underline{r} | \Psi \rangle$. Then $|\Psi(\underline{r})|^2$ is a quantity analogous to P_1 from Eq. (1.9) (see Section 1.3). Precisely formulated, not directly $|\Psi(\underline{r})|^2$ but instead $|\Psi(\underline{r})|^2 \, \mathrm{d}V$ is the probability to find the quantum object with a position measurement in a small volume $\mathrm{d}V$ around the position $\underline{r}$. Since the position is a continuous quantity, we have to refer to the volume element $\mathrm{d}V$. This is unnecessary for discrete measurement values.

Finally, the problem remains that we have to introduce the dynamics, or more precisely the different dynamics. In the force-free case or under the influence of external forces, which are taken into account with the corresponding potentials, the state of the system changes continuously between two times t_1 and t_2:

$$|\Psi(t_1)\rangle \rightarrow |\Psi(t_2)\rangle \qquad (1.12)$$

Based on Fig. 1.8 this change would be illustrated by a continuous rotation of the vector $|\Psi\rangle$ with a time-dependent angle $\alpha(t)$. This evolution in time between the measurements is deterministic and causal. We are going to call it *dynamics I*. The differential equation that describes the evolution in time (1.12) in detail is the *Schrödinger equation*.

One important property of dynamics I should be mentioned here. Usually it is summarized in the following statement: "Quantum theory is linear". We ask ourselves how different states evolve under an identical influence from outside. *Linear* means that the coefficients a and b in Eq. (1.7) do not change during the dynamic evolution of the superposition. The evolution $|u(t_1)\rangle \rightarrow |u(t_2)\rangle$ and $|v(t_1)\rangle \rightarrow |v(t_2)\rangle$ implies the evolution

$$a\,|u(t_1)\rangle + b\,|v(t_1)\rangle \rightarrow a\,|u(t_2)\rangle + b\,|v(t_2)\rangle \qquad (1.13)$$

of the superposition.

As a consequence of the *measurement*, the state is also transferred to a new state, but in a very different way. When we take the above example, we have $|\Psi\rangle \rightarrow |\Psi_1\rangle$ when result 1 is measured. In Fig. 1.8 this means that the state vector "jumps". We have seen before that this is a nondeterministic dynamics, since we are not able to predict before the measurement, in which one of the eigenstates $|\Psi_1\rangle$ or $|\Psi_2\rangle$ the initial state will jump. We are going to call this dynamics, which is present in the measurement, *dynamics II*. The probabilities P_1 and P_2 are the absolute squares of the projections of $|\Psi\rangle$ on the orthogonal vectors $|\Psi_1\rangle$ and $|\Psi_2\rangle$. The measurement is therefore sometimes also called a *projection measurement* or *state reduction*.[14]

---

14) In order to restrict the formalism to the essential, we have not introduced the concept of observables (operators that represent measurement quantities). Since they may be characterized by giving the vectors in which a state is transferred with the corresponding measurement, the knowledge of $|\Psi_1\rangle$ and $|\Psi_2\rangle$ is sufficient for our considerations. The uncertainty relation (1.6) could be deduced from the fact that position and momentum operators do not commute.

The fact that we need to postulate two dynamics for the quantum theory, one of them deterministic, the other nondeterministic, is certainly extremely unsatisfying. The search for unification is therefore another current research program. The attempts to formulate a satisfying *theory of the quantum measurement* start from dynamics I.

1.7 The quantum Zeno effect: How to stop the dynamical evolution

As a simple application, we study a dynamical process evolving freely or under the influence of outer forces (dynamics I), which is interrupted again and again by a measurement of the same type (dynamics II). It shows then as consequence of dynamics II an amazing quantum-mechanical effect for dynamics II: the dynamical evolution according to dynamics I can be completely suppressed by repeated measurements of a quantum system. The quantum system is "frozen" in its initial state. This effect is named the *quantum Zeno effect* in memory of the Greek philosopher Zeno (490–430 BC), who formulated a paradox, according to which any movement should be logically impossible, the so-called "paradox of Achilles and the tortoise". The infinitesimal calculus makes it possible to solve Zeno's paradox. A Zeno effect is impossible in classical mechanics. In contrast to this, the quantum Zeno effect is no paradox. It can be experimentally demonstrated (Itano et al. (1990)) and understood as a direct consequence of the peculiarities of the quantum measurement process.

Let us assume that we are measuring again a physical quantity, where the states $|\Psi_1\rangle$ and $|\Psi_2\rangle$ belong to the two measurement results. We further assume that the quantum system is in the initial state $|\Psi_2\rangle$ at an initial time t_0 as a result of a measurement. Under some influences from outside, which can for example be given by potentials, the system evolves according to dynamics I. This means for the state vector $|\Psi(t)\rangle$ that it is slowly rotating away from position $|\Psi_2\rangle$ with a time-dependent angle $\alpha(t)$ (see Fig. 1.8). After the next measurement, the probability to find the system either in the state $|\Psi_1\rangle$ or in $|\Psi_2\rangle$ is given by the square of the projection of $|\Psi(t)\rangle$ on these states. When the measurement is repeated after a very short time, the angle α is still very small and the probability for the state to be projected back to the state $|\Psi_2\rangle$ is much larger than the probability to "jump" to the state $|\Psi_1\rangle$. When the first case happens, the evolution therefore starts again with the state $|\Psi_2\rangle$. When the measurement is made once more after a very short time, the above applies again. In total, the probability to find the system still in the state $|\Psi_2\rangle$ after a very fast sequence of measurements of the same kind is high. The experiment confirms this.

At least theoretically we can consider now the limiting case that the time intervals approach zero. In this case one can show that the dynamical evolution is completely prevented and the system remains in its initial state $|\Psi_2\rangle$. This, remarkably, also holds when the initial state is not an eigenstate of the corresponding measurement quantity. The theoretical reason for this is quite simple. However, this is beyond the scope of this discussion. Strictly speaking, the limiting case cannot be realized due to the final duration of each measurement process. Investigations about the existence of exceptions, where special dynamics of type II prevent the quantum Zeno effect, are a topic of current research.

1.8 Photons: quantum objects of light

Once again we return to electrodynamics and ask ourselves if there might be any structure behind the electromagnetic fields, which is similar to the structure of quantum mechanics. This is indeed the case. First, we take a look at the particle aspect. For the objects of quantum mechanics this aspect became apparent in the always well-defined quantities of mass and charge. Electromagnetic fields do not have these two properties. A quantum-type structure manifests itself in a different way.

The photoelectric effect shows that energy can only be taken from a light field in portions of size $E = h\nu$. ν is the frequency of light and h is Planck's constant. This holds for all electromagnetic fields correspondingly. Atoms can be accelerated or decelerated with laser light. When the experiments are analyzed in detail, one finds that the momentum transfer is also a quantum-type exchange in "packages" of size $p = h\nu/c$. These mass- and charge-free exchanged energy-momentum packages are named *photons*. They move with the speed of light c, and they are a new type of quantum object. All effects in quantum optics are based on this quantum nature of light.

The electromagnetic waves and diffraction phenomena are now explained in the same way as for the material quantum objects: the description of Section 1.6 can be adopted correspondingly. Therefore our quantum domain is extended. Again we are dealing with many photons prepared in the same way. In special cases the current of photons can be thinned out, so that only one photon is inside the experimental setup.

Electromagnetic waves oscillate perpendicularly to their propagation direction. When we look at this linear polarization of the photons, the analogy to the formalism described above for material quantum objects becomes particularly clear. Light that is propagating in the z-direction can be polarized for example in x-direction by means of a polarization filter (polarizer). When this light falls on a second polarization filter (analyzer) with a perpendicular orientation in the y-direction, no light passes through. When the polarizer is turned by $45°$ into a diagonal orientation, only half the intensity will come through an analyzer aligned in the x-direction and also only half of the intensity will come through the analyzer in the y-direction. In the general case, the polarization is turned by an angle α against the x-direction. The intensities for both analyzer directions result from exactly the same rule that we established for the probabilities in Section 1.6. Again we get to the ensemble experience described in Section 1.3. In summary, we can conclude that for the description of photons a quantum state "linear polarization" has to be introduced, to which the same rules apply as the ones that we have already studied for the special case of our quantum state behind the double slit.

1.9 Is it possible to see in the dark? Interaction-free quantum measurements

Of course it is impossible to see in the dark. A photon must be scattered by an object so that one can see it — but this photon, of course, is not present in absolute darkness. However, we can reformulate the question: is it possible to prove the presence of an object at a defined position without the object being hit even by a single photon? We are going to demonstrate that it is indeed possible. This can already be shown with very simple considerations, which we are going to discuss first (Elitzur and Vaidman (1993)). In the second part of this section, this effect will serve as a simple application of the quantum-theoretical formalism that we discussed in Section 1.6.

A Mach–Zehnder interferometer is shown in Fig. 1.9. It consists of a light source (star), two semipermeable beam splitters A and B, two ideal mirrors K and L as well as the two detectors D_d and D_h. The paths 1 and 2 have exactly the same length. Light should be transmitted and reflected by the beam splitter to the same extent. With each reflection, either by the beam splitter or by the mirror, the phase of the electromagnetic wave is changed by $\pi/2$. In contrast, no phase shift occurs when the wave passes through the beam splitter. The "rules of the game" for the interferometer are defined by this characterization of the optical elements. Interference occurs on each way from B to one of the two detectors.

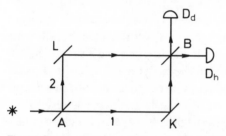

Fig. 1.9 Schematic representation of a Mach–Zehnder interferometer consisting of a light source (star), two semipermeable beam splitters A and B, two ideal mirrors K and L and the two detectors D_d and D_h.

At first we shall consider classical electromagnetic waves. When following the sequence of phase shifts, one realizes that two waves with different phase shifts run from B towards the detector D_d: one wave with phase shift $3\,\pi/2$ coming along path 2 and another with phase shift $\pi/2$ coming along path 1. Both waves are superimposed behind B and due to a resulting phase shift of π they cancel out each other. Therefore, the detector D_d is not responding, while the detector D_h is, as can be shown by analogous considerations.

We pointed out above that light consists of single photons. Therefore, it is quantum-theoretically possible and feasible that there is only one single photon within the interferometer. This nonclassical situation leads to new effects for the optical interferometer. Similar to the case of the double slit, no quantum object can be found in areas where the classical waves cancel each other by interference. According to this, the single photon in the case discussed here is registered always only in detector D_h but never in detector D_d.

Now we are going to insert an object into path 2, which is either absorbing or scattering a striking photon by an interaction, so that the photon can no longer reach the beam splitter B. In the same way as before, only one single photon enters the interferometer. We give a very simplified description of what happens in terms of paths and turn to a more correct one afterwards. At beam splitter A it is reflected to path 2 with a probability of $1/2$ and hits the object. Also with a probability of $1/2$ it enters path 1, where it reaches the beam splitter B. From there, it goes through to the detector D_d or the detector D_h with a total probability of $1/2 \times 1/2 = 1/4$. So having an object in beam path 2, a photon is registered in detector D_d in one quarter of all cases. Without the object, no photon is observed at all in detector D_d. Therefore it is clear: we only register a photon in detector D_d when there is an object in beam path 2. Since only one single photon has been used and this one was registered in detector D_d, no interaction of the photon with the object could have occurred, because in such a case the photon would not have reached one of the detectors.

To illustrate this, we are going to discuss the following practical example. Let us assume that the fuse of a bomb, which ought to be deactivated, is so sensitive that the explosion would be triggered by the impact of one single photon. It is impossible to search for such a bomb with classical light. With the arrangement described above there is, however, not too bad a chance — namely a probability of $1/4$ — to find the bomb without igniting it. The fact that once again it is only possible to make a statement about a probability, is characteristic of quantum-theoretical effects. In situations where such a prediction is "better than nothing", the use of quantum effects opens up unexpected technical possibilities. Quantum computers are another application of this.[15]

The interaction-free quantum measurement is a simple example, which can be used to test for photons the formalism of quantum theory from section 1.6. The state that describes a photon running in

---

15) Quantum computer and quantum information theory are discussed in more detail in the chapters of R. F. Werner (Chapter 7) and H. Weinfurter (Chapter 6) of this book.

the x-direction is marked by $|x\rangle$; for the y-direction we use $|y\rangle$ correspondingly. With a reflection at the mirror or the beam splitter, a phase jump of $\pi/2$ occurs. This means for the corresponding state a multiplication with i because of the phase factor $\exp(i\,\pi/2) = i$. Therefore, the mirrors in L and K cause the following state changes:

$$|x\rangle \to i\,|y\rangle\,, \quad |y\rangle \to i\,|x\rangle\,. \tag{1.14}$$

The beam splitters A and B transfer the state to a superposition, where the phase jump for the reflected outgoing state has to be taken into consideration:

$$|x\rangle \to \frac{1}{\sqrt{2}}\left(|x\rangle + i\,|y\rangle\right) \tag{1.15}$$

$$|y\rangle \to \frac{1}{\sqrt{2}}\left(|y\rangle + i\,|x\rangle\right) \tag{1.16}$$

These are our quantum-theoretical "rules of the game" for the state modifications by the optical elements.

Now we follow again a photon that enters with the state $|x\rangle$. The complete evolution of the state, when there is no scattering or absorbing object in one beam path, is given by the following sequence of state transitions:

$$|x\rangle \to \frac{1}{\sqrt{2}}\left(|x\rangle + i\,|y\rangle\right) \to \frac{1}{\sqrt{2}}\left(i\,|y\rangle - |x\rangle\right) \to$$

$$\to \frac{1}{2}\left(i\,|y\rangle - |x\rangle\right) - \frac{1}{2}\left(|x\rangle + i\,|y\rangle\right) = -\,|x\rangle \tag{1.17}$$

The first arrow describes the transition at the beam splitter A. After the reflection at the mirrors L and K, the state after the second arrow is present. The effect of the beam splitter B is illustrated with the third arrow. This beam splitter, according to Eqs. (1.15) and (1.16), causes the transition to further superpositions, which are added up on the right side of the equation. After the beam splitter, our photon is thus in the state $-|x\rangle$ and it arrives at the detector D_{h}. The detector D_{d} therefore never responds in this configuration.

Now we imagine having a scattering or absorbing object inserted in the path between L and B. The state $|x\rangle$ that is running towards this object is transferred to the photon state $|s\rangle$ by scattering or absorption. When we take a look at the state evolution, Eq. (1.17) has to be modified accordingly:

$$|x\rangle \rightarrow \frac{1}{\sqrt{2}}\left(|x\rangle + i\,|y\rangle\right) \rightarrow \frac{1}{\sqrt{2}}\left(i\,|y\rangle - |x\rangle\right) \rightarrow$$

$$\rightarrow \frac{1}{\sqrt{2}}\left(i\,|y\rangle - |s\rangle\right) \rightarrow \frac{1}{2}\left(i\,|y\rangle - |x\rangle\right) - \frac{1}{\sqrt{2}}\,|s\rangle \qquad (1.18)$$

The last state is present after the influence of the beam splitter B. The probabilities are obtained as usual from the squares of the absolute of the prefactors. The probability of the response of detector D_h or detector D_d is $1/4$ in each case. The probability for a photon to be scattered or absorbed by an object is $1/2$. It is crucial that after inserting the object, the state $|y\rangle$ also appears in the resulting superposition of Eq. (1.18), in contrast to Eq. (1.17). This means that the detector D_d can now respond, which was impossible without the object. When the detector actually responds, we obtain the information that an object has been in path 2. In the first case, the single photon that was used to carry out our experiment has arrived unharmed in detector D_d, neither being scattered nor absorbed. In this sense, we were able to prove the existence of an object without having any interaction.[16] An experimental realization can be found in Kwiat et al. (1995).

1.10 What is real? Interpretations of quantum theory

When in classical mechanics or quantum mechanics the state of a system can be specified, one has the greatest possible knowledge about this system in the following sense: A prediction can be made for all possible measurements on this system. In quantum mechanics this prediction is only statistical but the expectation value (average value) and the variance are well defined. We have already seen that the quantum-mechanical process of measuring, as opposed to the classical measurement, alters the state. Quantum mechanics is for equally prepared states *non deterministic* in respect of the results, which are obtained in a specific measurement on these states. It is *acausal* (already because of dynamics II). The quantum probabilities are primary

---

16) Although generally used, the denotation "interaction-free" is not entirely appropriate. The third arrow in Eq. (1.18) represents an interaction. For a more detailed analysis see Audretsch (2005).

and fundamental, and real chance accordingly exists. Even if we ensure that there is always only one object within the experimental setup, the ensemble result for the double slit is different from the superposition of the results from two single slits. Oversimplified, one might say: the single object "notices" the presence of both slits. In this sense, quantum mechanics and quantum physics are generally *nonlocal*.

From classical mechanics, we know for flying tennis balls how to imagine the states behind the slits. But what is there in quantum physics? Obviously this leads to the question about the interpretation of quantum theory. First, we should clarify what we understand by an interpretation, because this term is not uniformly used.[17]

A physical theory is on the one hand a system of mathematical symbols with rules applied to them, which are typically used to derive the results from basic equations. This is the *syntax*. The advantage of a mathematical syntax for a physical theory is that it makes it easier to track down errors and inconsistencies. The ways to draw conclusions become intersubjectively compelling and in this sense, an objectivity of the physical knowledge is created.

Of course, a physical theory has to be more than just a mathematical theory: it should make statements about a part of reality. Therefore, there are *mappings* between some of the mathematical symbols on one side and objects in reality on the other side. For example, r denotes the space of a classical body, m its mass, and t the time on a clock. We need, as an essential core part of a physical theory, mapping rules or *correspondence rules* that link certain mathematical quantities with the pointer positions of measuring instruments, and thus with the results of the measurements. Whether we can say about the numerous other mathematical quantities that appear in the physical theory that they are related to something in reality and thus have a physical meaning (*semantics*) remains an open question.

In classical physics it is no different. In electrodynamics there is a mathematical quantity $\underline{E}(\underline{r}, t)$ that is named the electric field. This is

---

17) See also the chapters of C. Held (Chapter 3) and M. Esfeld (Chapter 10) in this book. There are many discussions of conceptional problems of quantum theory. Such, with a close reference to the theoretical formalism, can be found for example in Primas (1981), Readhead (1987), Mittelstaedt (1989), Omnès (1994), Peres (1995), Home (1997), Mittelstaedt (1998), Espagnat (1999) and Auletta (2000). For a review, see also Audretsch and Mainzer (1990/1996) and Audretsch (2005).

a *theoretical term*, because primarily we are only observing the mechanical behavior of charges. There is the option to regard $\underline{E}(\underline{r}, t)$ and the differential equations for $\underline{E}(\underline{r}, t)$ as purely mathematical auxiliary quantities and relations that allow us to deduce statements that can be related to measurement results. These, for example, can be statements about the dynamical behavior of charged spheres. The physical "existence" of electric fields then would not be assumed. With this kind of an approach to physical theories, the only goal of the theory would be the prediction of experimental results. The one who chooses this position negates the existence of an objective reality that is independent of what observers are recording. But one might also aim for an understanding of the physical world by means of a theory. In that case, one would say $\underline{E}(\underline{r}, t)$ is actually describing something real. There is an element of reality described by $\underline{E}(\underline{r}, t)$, which is called the electric field. This approach is common among physicists, but it is important to realize that it is not necessary to share it when the purpose of a physical theory is solely seen in the prediction of measurement results.[18]

What is considered to be *real* is obviously dependent on how a theory is interpreted. Already classical physics is usually given an *interpretation* beyond the correspondence rules of the essential core. These interpretations go beyond the nonreducible core statements, which relate directly some terms of the theory with measurement results. However, different interpretations can be attributed to the same experimental consequences: they cannot be distinguished experimentally. They cannot be falsified. This situation becomes even more complex in quantum mechanics.

The admission of different interpretations, though, must not be mistaken with the proposal of a theory for the quantum domain which differs from quantum theory but nevertheless leads for all experiments to the correct data. This would then not be a new interpretation of an old theory, but a truly new competitive theory. It is conceivable that there are experiments for which this theory predicts results other than quantum theory. In such a case, a decision

---

18) The many-worlds interpretation (Everett (1957)) is usually regarded as being more than an interpretation in the above sense. Theoretical problems in connection with the quantum process are solved in a very speculative way in which finally a connection with the state of mind of the observer is established. With this, a theory is asserted, which goes far beyond the present range of application of quantum theory.

between both theories can be forced by an experiment. We will come to an example in Sections 2.4 and 2.6.[19]

The interpretations, however, are on the other hand not a part of philosophy by themselves. They still belong together with the physical theory. It is one of the aims of a physical theory to lead beyond the prediction of measurement results to a conception about the physical world. Interpretations are in this sense physics but not metaphysics. Nevertheless, they have philosophic or metaphysical implications. On these, natural philosophy typically sets in.

In quantum mechanics, interpretations also try to give an answer to the question: What is *real*? What, based on the theory, can be said about reality beyond the prediction of measurement results? As a starting point one could ask: What is the state vector? It describes a state — but of what?

The interpretation that one might name the *Copenhagen Interpretation*, because ideas from the early days of quantum mechanics are assimilated within, is today only of historical relevance.[20] A phrase from Nils Bohr is placed at its center: "There is no quantum world."[21] There is no quantum world and no quantum objects. Only the phenomena are real. "Behind it" there is nothing. The quantum world is a mental construct. The state vector is a purely mathematical auxiliary quantity without correspondence in reality. It serves for the calculation of probabilities of macroscopic events, for example of measurement results. The "quantum object" is nothing but a manner of speaking that facilitates the communication about a computational procedure. A term like "electron" is hence just a practical abbreviation that is referring to a whole complex of calculations. The measurement instruments are classical devices, which are not to be described quantum mechanically. The calculations are finally just providing statements about the classical states of the measurement instruments. The complementarity of position and momentum in the way it is shown by the uncertainty relation has its origin in the fact that no measurement instrument exists for a combined measurement of position and momentum. For the supporters of this extremely pragmatic and

---

19) 2 refers to the article "View into the quantum world II" hereinafter. 2.4 indicates the Section 2.4 in there.

20) The addition "Copenhagen" is so dazzling and ambiguous that strictly speaking it should be replaced. A description of the historical situation can be found in the following article by C. Held.

21) Quoted according to Primas (1981), page 101.

---

minimalistic interpretation, the two challenges "establish dynamics II from dynamics I" and "trace the behavior of classical objects back to quantum mechanics" are completely irrelevant. This shows very clearly that interpretations can definitely be of great consequence for the design of research programs and for the motivation of scientists.

In contrast to this interpretation, the two problems set out above are absolutely meaningful from the point of view of an *ensemble interpretation*. In this case, the quantum world exists. The statements of quantum mechanics and with it the state vector $|\Psi\rangle$, refer to a statistical ensemble of infinitely many systems, all prepared in the same way. Therefore, statements are never made about a single quantum object. The relative frequency of measurement results can be predicted using the state vector $|\Psi\rangle$. Since in practice only finite numbers of systems are available, this represents an approximation. With this interpretation one remains entirely on the deterministic level. The reality represented by $|\Psi\rangle$ is precisely a matter of the entirety of measurements already performed. Again, there is a strong limitation on statements about experimental data. Therefore, this is often called a *minimal interpretation*. It contains all that the present-day physicists can agree on without problems. For the example discussed above, it is the interference pattern on the screen. A single object within the experimental apparatus, for example at the double slit, has no counterpart in the theory. Nothing is stated about its reality. It is questionable whether it appears to be meaningful within this framework to carry out the transition from quantum objects to large molecules, biological systems and classical objects.

Nowadays, single atoms and ions can be stored and manipulated in traps. The reality of these objects is generally assumed beyond question. Therefore, they should be represented. This is the common opinion among physicists today. In this *single-system interpretation*, the state vector $|\Psi\rangle$ now directly refers to real single objects and their attributes. The single objects, besides, are mostly of a microphysical nature. In this interpretation, the measurement results do not appear as the primary *references* of theory anylonger, but instead they are the single objects on which the measurements are performed. The quantum object really exists also before and after the measurement. A justification for this interpretation should be searched for by the fact that properties like mass, charge and size of the spin of a quantum object always have the same value, independent of the preparation of the quantum system and the performed experiments. The quantum object by itself "has" these *classical properties*. They do not rely on "relations" between the system and the apparatus for its preparation or its measurement. These properties therefore can be specified as being *objective* and *real*.[22] The quantum system obtains an "existence" in between preparation and measurement. The state refers now to the preparation procedure. At the same time, the question about the nonclassical attributes becomes important.

A measurement transfers the object to a state that corresponds to the measured value. When the same measurement is immediately repeated, the same measured value results. For example, the repetition of a position measurement finds the object in the same position. The position uncertainty of the object after the first position measurement is zero. When a reliable prediction for the result of the measurement of a quantity (for example the position) can be made for a quantum state, one would say that this property (of having a position) must be assigned to the object in that state.

When a physical quantity is undetermined in a given state (for example the position if $\Delta x \neq 0$), the object does not have the attribute (of having a position); this attribute does not exist at all in this case. In the extreme case of a momentum eigenstate ($\Delta p_x = 0$), each measurement of the component p_x of the momentum gives a well-defined value but no prediction about the result of a position measurement is possible ($\Delta x = \infty$). Therefore, one would have to say that objects in the momentum eigenstate have the property of a momentum but the property of a position cannot be assigned to them. In the position measurement and thus under participation of the measuring instrument and in the course of a state modification, the object receives the property of having a position. A subsequent momentum measurement not only destroys the value of the property, but the property itself. As an expression of the uncertainty relation, the object will in general be in a state in which *neither* the attribute position *nor* the attribute momentum can be assigned in this sense. Position and momentum are then *potential* properties. Not until the measurement are they *actualized*.

---

22) To deduce the reality of atoms from the reality of the apparatuses that are used for the preparation and detection takes a great formal effort. Whoever would like to gain some insight may consult the book by Ludwig (1985).

Such *neither-nor-objects* are unknown from our everyday surroundings. However, this kind of object can be drawn, as in Fig. 1.10:
it consists neither of three tubes nor of two boxes. As a matter of
fact, when watching exclusively the upper part one can see three
tubes, when watching the lower part there are two boxes. Only as
the result of a specific measurement "look at the top" does the object
get the attribute to consist of three tubes. This cannot be assigned
before. It is only a potential attribute inherent to the object. This is
equivalent for the two boxes. Both attributes mutually exclude each
other. When the measurement "look at the middle" is made, we can
learn neither about the attribute "tubes" nor about the attribute
"boxes".

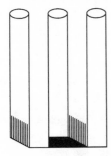

Fig. 1.10 A neither-nor-object.

When the comparison with classical physics is made, one realizes that not only the concept of "causality" but also the concepts
"attribute" or "property" have become a great deal weaker, more
general and more flexible. This reflects once again the fact that quantum theory is the more general and more comprehensive theory.
It requires therefore — compared to the everyday physics-oriented
common speech which is everyday physics-oriented — a *reduced level
of speach*. When talking about quantum objects and their behavior we
must not suggest, for example, that objects are particles or waves or
that they always have all properties at the same time, by inconsider-
ately using common speech formulations.

Finally, it should be pointed out that the concept of "information"
plays a completely new role compared to classical physics. There, information obtained by measurement tells us what has already existed
before and is still there after the measurement, whereas in quantum

physics the single measurement yields only information about the state after the measurement. In this sense, the quantum theory is of course more general but the statements are also weaker. This has consequences that will be discussed in Section 2.7. In reverse, a state $|\Psi\rangle$, from Fig. 1.10, which can assume all orientations α, can obviously be used to store more information than two bits. When these two peculiarities of quantum physics are combined with the entanglement of states described in the following chapter, an information theory of a completely unusual kind unfolds, with a wealth of new possibilities for information processing and transmission. This quantum information theory applies also to quantum computers that are constructed using entangled quantum systems. A fast-developing section of quantum physics has emerged in this direction over the past few years.[23]

References

- M. Arndt et al. (1999), *Nature* **401**, 680.
- J. Audretsch, K. Mainzer (eds.) (1990 and 1996), „Wieviele Leben hat Schrödingers Katze? Zur Physik und Philosophie der Quantenmechanik", Heidelberg.
- J. Audretsch (2005), „Verschränkte Systeme — die Quantenphysik auf neuen Wegen", Wiley-VCH, Weinheim. An English translation of the textbook with the title "Entangled Systems" will be published in 2006 by Wiley-VCH, Weinheim.
- G. Auletta (2000), "Foundations and Interpretation of Quantum Mechanics", World Scientific, Singapore.
- P. R. Berman (ed.) (1997), "Atom Interferometry", Acad. Press, San Diego.
- A. C. Elitzur, L. Vaidman (1993), *Found. Phys.* **23**, 987.
- B. d'Espagnat (1999), "Conceptual Foundations of Quantum Mechanics" (2nd edn.), Perseus Books, Reading.
- H. Everett (1957), *Rev. Mod. Phys.* **29**, 454.
- R. Feynman, R. Leighton, M. Sands (1965), "The Feynman Lectures on Physics", Volume III, Addison-Wesley, Reading.

---

23) Details can be found in the articles from R. Löw, T. Pfau, H. Weinfurter and R. F. Werner in this book

- D. Home (1997), "Conceptual Foundations of Quantum Physics: An Overview from Modern Perspectives", Plenum Press, New York.
- W. M. Itano, D. J. Heinzen, J. J. Bollinger, D. J. Wineland (1990), *Phys. Rev.* **A41**, 2295.
- P. G. Kwiat, H. Weinfurter, T. Herzog, A. Zeilinger, M. A. Kasevich (1995), *Phys. Rev. Lett.* **74**, 4763.
- G. Ludwig (1985), "An Axiomatic Basis for Quantum Mechanics", Springer-Verlag, Berlin.
- C. Macchiavello, G. M. Palma, A. Zeilinger (2000), "Quantum Computation and Quantum Information Theory", World Scientific, Singapore.
- S. F. Mason (1953), "A History of the Sciences", p. 391–392, Routledge & Kegan Paul LTD, London, 1953.
- P. Mittelstaedt (1989), „Philosophische Probleme der modernen Physik" (7th edn.), Bibliographisches Institut, Mannheim.
- P. Mittelstaedt (1998), "The Interpretation of Quantum Mechanics and the Measurement Process", Cambridge University Press, Cambridge.
- R. Omnès (1994), "The Interpretation of Quantum Mechanics", Princeton University Press, Princeton.
- A. Peres (1995), "Quantum Theory: Concepts and Methods", Cluver, Dordrecht.
- H. Primas (1981), "Chemistry, Quantum Mechanics and Reductionism", Springer-Verlag, Berlin.
- H. Rauch, S. A. Werner (2000), "Neutron Interferometry: Lessons in Experimental Quantum Mechanics", Clarendon Press, Oxford.
- M. Readhead (1987), "Incompleteness, Nonlocality and Realism", Clarendon Press, Oxford.
- A. Tonomura (1998), "The Quantum World Unrevealed by Electron Waves", World Scientific, Singapore.
- J. A. Wheeler, W. H. Zurek (1983), "Quantum Theory and Measurement", Princeton University Press, Princeton.

2

View into the quantum world II: entanglement and its consequences

Jürgen Audretsch

2.1 Introduction

At the center of the second part of our "view into the quantum world" are composite quantum systems. Their states may be entangled. This is the basis of a wealth of new, typical quantum-physical phenomena. We are going to select a few of these phenomena, which are of fundamental interest as well as revealing possible applications for the future.

In Section 2.2, the concept of entanglement is introduced and the EPR correlations linked to it are described. In Section 2.3 we give a first application concerning the which-path information, which refers again to the double-slit experiment. A fundamental experiment, based on how many of the peculiarities of entanglement can be clarified, is the experiment with two entangled photons of which the polarization is measured in separate places. We will describe this in Section 2.4 in the version of the Orsay experiment and discuss the experimental answer to the question, if quantum theory, using hidden variables, can be understood as a local realistic theory, and thus as a part of classical physics. Bell's inequality deduced in Section 2.5 will prevent that. Based on the same experimental setup, the procedure used in quantum cryptography is illustrated in Section 2.6. The no-cloning theorem of Section 2.7 once again puts the fundamental difference between quantum physics and classical physics into concrete terms. Its application to the Orsay experiment clarifies in which way quantum theory protects itself from the possibility of getting into contradiction with the theory of special relativity. The important consequences

Entangled World, Jürgen Audretsch
Copyright © 2005 WILEY-VCH Verlag GmbH & Co. KGaA, Weinheim
ISBN: 3-527-40470-8

of the existence of entangled systems for a philosophic analysis of the quantum world are addressed in Section 2.8. This concludes the second part of the introduction to the new results and the current problems of quantum physics.

2.2 Compound systems and entangled quantum states

At first we will extend the vector formalism that we got to know in Section 1.6.[1] In the following, we will look at a quantum system, which is *composed* of two quantum systems of the type previously discussed. We are going to call them the *subsystems* 1 and 2 and denote the state vector of the *composed system* by $|\Phi\rangle$. The subsystems could be, for example, two photons flying away in opposite directions, an atom and a photon, etc. For more than two systems one should proceed accordingly. Different degrees of freedom of a quantum object may in general, be conceived as such subsystems. Therefore, this discussion could also be about the degree of freedom of the path and the degree of freedom of the spin of an atom. A precondition for all cases is the possibility to carry out separate measurements on the single subsystem. As an example, we will consider two photons separated in space from each other and measure their states of polarization. We are going to formulate the theoretical description at first generally, starting from Section 1.6.

We restrict ourselves to the special case that any measurement on the subsystems can have only two results, called 1 and 2 for system 1 (see Section 1.6) and u and v for system 2. In the following, we are interested in *double measurements* and the corresponding *pairs of results*. For this purpose we transfer the description of the measuring process (dynamics II): system 1, after being measured is — depending on the result — either in the related eigenstate $|1_1\rangle$ or $|2_1\rangle$, while system 2 is in the state $|u_2\rangle$ or $|v_2\rangle$, correspondingly. The index marks the system that was measured. The core symbol marks the measurement result and the associated state vector. Similar to Section 1.6, all state vectors are again pairwise normalized and orthogonal to each other: $\langle 1_1 | 2_1 \rangle = 0$, $\langle u_2, v_2 \rangle = 0$.

---

1) 1 refers to the previous chapter "view into the quantum world I". Section 1.6 therefore marks Section 6 of that chapter.

The following combinations of measurement results are possible: $(1, u)$, $(1, v)$, $(2, u)$, $(2, v)$. For a certain state $|\Phi\rangle$ of the composite systems, the different combinations always occur with the same fixed frequencies. For a single pair of measurements, however, it is still totally random which one of the combinations is measured. Each one of the two single systems is in a well-defined state after the double measurement. In the case of measurement result $(1, u)$ for example, these are the states $|1_1\rangle$ and $|u_2\rangle$. We mark this particular state of the compound systems with $|1_1, u_2\rangle$. Sometimes this is also written as $|1, u\rangle$ or $|1\rangle|u\rangle$.

The general state of a composed system consisting of two subsystems is again obtained by *superposition* of the states related to the possible measurement results:

$$|\Phi\rangle = c(1, u)\,|1_1, u_2\rangle + c(1, v)\,|1_1, v_2\rangle$$
$$+ c(2, u)\,|2_1, u_2\rangle + c(2, v)\,|2_1, v_2\rangle \qquad (2.1)$$

The prefactors $c(1, u)$, $c(1, v)$, etc., are complex numbers. Inner products are formed "by the subsystem". The results are multiplied as for example $\langle 1_1, u_2 | a_1, b_2 \rangle = \langle 1_1 | a_1 \rangle \langle u_2 | b_2 \rangle$.

The known rules for the dynamics II are valid. When a *measurement* at the first subsystem gives the result 1, this system is transferred into the state $|1_1\rangle$ and the entire system is changed to the state

$$|\Phi\rangle \rightarrow |\Phi'\rangle = c'(1, u)\,|1_1, u_2\rangle + c'(1, v)\,|1_1, v_2\rangle$$
$$= |1_1\rangle\,(c'(1, u)\,|u_2\rangle + c'(1, v)\,|v_2\rangle) \qquad (2.2)$$

accordingly.[2] In the second equation we have applied the product rule. The state $|\Phi\rangle$ is normalized: $\langle \Phi | \Phi \rangle = 1$. The probability to find for the state $|\Phi\rangle$ the pair of measurement results $(1, v)$ is again given by $|c(1, v)|^2$. The probability is also in this case a statement about the relative frequency of measurement results from many measurements, which were made at systems in the same state $|\Phi\rangle$. A statement about the succession of the measurements at systems 1 and 2 is not required.

---

2) Due to the fact that $|\Phi'\rangle$ is normalized, the c' are obtained from the corresponding c of Eq. (2.1) by adding a common normalization factor. The sum over all probabilities equals one.

When system 1 is in the state $|a\rangle$ and system 2 is in the state $|b\rangle$, the state

$$|\Omega\rangle = |a_1, b_2\rangle \qquad (2.3)$$

is assigned to the compound system. This is a special case of Eq. (2.1) and is called the *product state* or direct product. We have already seen for Eq. (2.2) that measurements at a subsystem are transferring to product states of the entire system.

It is essential now that among the states $|\Phi\rangle$ of Eq. (2.1) there are states that cannot be written as a product state. We discuss as a simple example the nonfactorizable sum:

$$|\Phi\rangle = \frac{1}{\sqrt{2}} \left(|1_1, u_2\rangle + |2_1, v_2\rangle \right) \qquad (2.4)$$

For this quantum state one can see that a measurement at system 1 gives in half of all cases the result 1 or 2, respectively. Correspondingly, one gets for measurements at system 2 in half of the cases the result u or v, respectively. But the probability to find a certain pair is not equal to the product of the probabilities for the single measurement results. The probabilities are not independent of each other, reflecting the fact that the state cannot be factorized. The probability for the pair $(1, v)$ is, for example, not $1/4$ but zero. This combination will never occur for the state $|\Phi\rangle$ of Eq. (2.4). When such circumstances for the occurrence of probabilities are present, one calls the state of the compound system an *entangled state*. We have already seen in Section 1.2 that the superposition is the central feature of the quantum theory. For the compound systems, the superposition of the product states leads to entangled states. The state $|\Phi\rangle$ in Eq. (2.4) is a typical representative. We will get to know even simpler examples. One can also say in this case, that the two subsystems are *entangled*.[3]

The property of entanglement is just not present for product states, as the one in Eq. (2.3). We take a look at the example $|\Omega\rangle = |1_1, v_2\rangle$. Here, the probabilities to measure the value 1 at system 1 and the value v at system 2 are 1, respectively, and the probability for the pair of results $(1, v)$ is also 1. Product states are not entangled states.

---

3) Erwin Schrödinger used this term in his influential works from 1935 for the first time (Schrödinger (1935)). In the same article, he also described the famous paradox of "Schrödingers cat" (see also the contributions in Audretsch and Mainzer (1990, 1996)).

For the experimental creation of entanglement see Sections 6.3 and 9.7.

Entangled states and product states show for measurements at the subsystems totally different frequencies for the different pairs of measurement results. In order to express this, one says about the entangled state that it is *EPR-correlated*.[4] For entangled states, the pairs of measurement results show *EPR-correlations*.

We still have to add the dynamical behavior of compound systems under the influence of potentials (Dynamics I, see Eq. (1.12)). An important observation is made there, which we can only report in this context without any justification: Let us assume that the system is initially in a product state. The compound system will remain in a product state only for an undisturbed evolution or when the influence from outside can be completely separated into two components, one of them affecting only system 1 and the other one system 2. This occurs exactly when the two subsystems are purely formally composed to an entire system; in this case they could also be handled as separate single systems, as was done in the previous chapters. Such a situation is the exception. In general, interactions and influences are present that cannot be split up in this way. Correspondingly they transfer as a rule an initial product state during the dynamical evolution to an entangled state (for example $|\Omega\rangle \rightarrow |\Phi\rangle$). The fact that systems are in entangled states is therefore the "standard case" in the quantum domain.

2.3 Which-path information: entanglement destroys the ability to interfere

Now we know of the concept of entanglement, we are going to take a look again at the experiment described in Section 1.5, where the atoms were illuminated directly behind the slit (Fig. 1.6). The flashing either behind slit 1 or behind slit 2 indicated through which slit the atom was flying, meaning which path it had taken. The consequence of this measurement was the disappearance of the interference image of the double slit. Strictly speaking, the complete quantum measurement process should be analyzed for the discussion of this effect. This, however, would go beyond the scope of this chapter. In-

---

4) EPR stands for the names A. Einstein, B. Podolsky, and N. Rosen (Einstein (1935)). For the historical background of this term see the following chapter by C. Held.

stead, we are going to demonstrate the role of entanglement with a slightly modified experiment. At the same time we can discuss the question, which is repeatedly asked in the popular science literature: do the waves coming from the single slit lose the ability to interfere only because the flash is registered in the human eye and then processed in the brain? Does it really matter that someone takes a note of the measurement result ("flashing behind aperture 1" or "flashing behind aperture 2")? Is therefore the human consciousness playing a decisive role in the quantum theory?

We are going to discuss a gedanken experiment (Scully et al. (1991)).[5] Atoms described by a plane Ψ-wave are again supposed to hit a double slit. Before the atoms actually reach the slits, they have to pass resonators that are installed separately from each other in front of each of the slits (Fig. 2.1). The resonators are empty cavities. When these cavities are correctly dimensioned, an appropriate atomic species is chosen and the velocity of the atoms is adjusted so that the suitable length of stay in the cavities results, the following can be observed: single atoms entering a cavity in the excited state emit exactly one photon into that cavity. In the course of this process they pass to the ground state and leave the cavity through the opening on the opposite side without being affected any further. The photon remains trapped in the cavity.

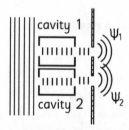

Fig. 2.1 Wave trains of the state function of an atom, which run through two cavities before reaching the double slit. There is always only one atom inside the setup. The previously excited atom emits a photon into one of the cavities and by doing so, it leaves a which-path information behind. Atom and photon are afterwards in an entangled state.

5) The way in which this kind of experiment can actually be realized in the laboratory is described in the chapter of G. Rempe (Chapter 5) in this book (cf. also (Kwiat et al. (1992))).

We consider again the situation that there is always only one single atom inside the whole installation. The excited atom then deposits a photon either in cavity 1 or in cavity 2, thus leaving behind a *which-path information* and is finally registered when hitting the screen. In this process, the experimentalist should by no means try to find out in which one of the previously empty cavities the photon was deposited. When looking again at the impact of many atoms on the screen, one realizes that the typical interference image for the double slit is lost. The registered image agrees as in Section 1.5 with the image that is formed when one of the slits was always blocked while the atoms were passing through the installation. The waves $\Psi_1(\underline{r})$ and $\Psi_2(\underline{r})$ were therefore obviously not interfering with each other.

The reasons for the fact that the generation of a which-path information and the interference are not compatible with each other can already be given with the restricted knowledge about entanglement available to us so far. The two cavities on one side and the atoms on the other are a compound system. The subsystem "cavities" is in the state $|1\rangle$ after the photon has been deposited in cavity 1 and in state $|2\rangle$ when the photon is in cavity 2. The vectors are normalized: $\langle 1|1\rangle = 1$, $\langle 2|2\rangle = 1$. The photon emitted by an atom must obviously be located either in cavity 1 or in cavity 2. The corresponding state vectors are accordingly orthogonal to each other: $\langle 1|2\rangle = 0$. Further on, the occurrence of the atomic state $\Psi_1(\underline{r})$ in a measurement is strictly linked to the cavity state $|1\rangle$, and correspondingly the state $\Psi_2(\underline{r})$ to $|2\rangle$. The composite system is therefore in the entangled state which is a superposition

$$\Phi(\underline{r}) = \frac{1}{\sqrt{2}} \left(\Psi_1(\underline{r}) |1\rangle + \Psi_2(\underline{r}) |2\rangle \right) \tag{2.5}$$

We are only interested in the subsystem "atom". The probability density for hitting point $\underline{r}$ on the screen results as

$$P(\underline{r}) = |\Phi(\underline{r})|^2 = \Phi^*(\underline{r})\,\Phi(\underline{r}) =$$
$$= \frac{1}{2} \left\{ |\Psi_1(\underline{r})|^2 \langle 1|1\rangle + |\Psi_2(\underline{r})|^2 \langle 2|2\rangle + \right. \tag{2.6}$$
$$\left. |\Psi_1^*(\underline{r})\Psi_2(\underline{r})|^2 \langle 1|2\rangle + |\Psi_2^*(\underline{r})\Psi_1(\underline{r})|^2 \langle 2|1\rangle \right\}$$

When the facts that the detector states are normalized and orthogonal to each other are taken into account, the vanishing of the last two terms follows. The comparison with the Eqs. (1.2) and (1.3) shows that the resulting probability density represents the situation in which no interference of the waves $\Psi_1(\underline{r})$ and $\Psi_2(\underline{r})$ occurs.[6]

The emission of the photon in one of the cavities is equivalent to leaving behind the which-path information, i. e. through which one of the cavities and thus through which one of the slits the atom was flying. This information is not readoff, but it could be readoff. The particle character of the quantum object is manifested in it. Connected to this is the disappearance of the interference image as the expression of the wave character in the theoretical description. For the subsystem "atom", the ability to interfere is already lost through the entanglement with the subsystem "cavities". Interfering paths become in principle distinguishable. Interference occurs only as long as the alternatives "going through slit 1" and "going through slit 2" are in principle indistinguishable.

For the discernibility it is therefore not necessary that a measurement instrument really determines in which cavity the photon is, thus reading off the which-path information stored in the system. Correspondingly, there is not needed that the level of information of a human observer is improved by reading such a measuring instrument. The loss of the ability to interfere can therefore not be traced back to the idea that the human conscience might play an essential role in quantum mechanics. Finally, it is also remarkable that the interference pattern on the screen disappears even though the two states $\Psi_1(\underline{r})$ and $\Psi_2(\underline{r})$ of the atom behind the double slit are not modified in the calculation above. Both of them are included in the superposition $\Phi(\underline{r})$. Further on, no reference to the uncertainty relation was necessary. The cause for the described effect should after all solely be seen in the entanglement of the two systems — cavities and atom — as it is reflected in the state (2.5).

In the following, we are going to illustrate the EPR-correlation further with a concrete example.

---

6) A theoretically satisfying analysis of this effect uses density operators. This, however, would go beyond the limits of this chapter.

2.4 Hidden variables: is it all just classical physics?

We take a look at the following experimental setup (cf. Fig. 2.2): appropriately excited calcium atoms pass to the ground state in a cascade of two steps. Since the lifetime of the middle level is extraordinarily short, this happens as if it is just a "single" process. During this process, an entangled photon pair (the composite system) is emitted. The single photons of the pair (the subsystems) have different frequencies and they fly away in the positive or negative z-direction, respectively. They reach the parallel aligned analyzers I and II, in which the polarizations in the x- and y-direction are measured. The distances between the analyzers and the source Q are of no relevance for the following considerations. They could also be different. One finds that only the measurement pairs (x, x) or (y, y) are registered, both with the same frequency. x and y mark the measured polarization direction in each case. The entangled state of this *2-photon system* therefore has the form

$$|\Phi\rangle = \frac{1}{\sqrt{2}} \left(|x_1, x_2\rangle + |y_1, y_2\rangle \right) \tag{2.7}$$

The photon flying towards analyzer 1 or 2 is marked here by the index 1 or 2, respectively. We will leave out the indices in the following. The reference to I or II follows from the succession.

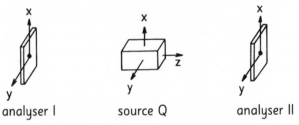

Fig. 2.2 Two entangled photons are generated in the source Q. The photons fly in negative or positive z-directions, towards the parallel-aligned analyzers in I or II, respectively, which measure their polarization direction.

The experiment described here is of fundamental importance for the discussion of the entanglement of compound quantum systems and the illustration of their properties. Based on experiences from previous experiments, this experiment was carried out by A. Aspect and coworkers in Orsay in 1982 (Aspect et al. (1981), Aspect et al. (1982)). It is therefore called the *Orsay experiment*. Nowadays there are also other possibilities to generate and to experiment with entangled photon pairs.[7]

Whenever a certain polarization direction (for example x) is measured for a single photon pair for example in I, the same direction is also measured in II (likewise x). This certain prediction can be made by an observer in I independent of the distance between I and II. A strong correlation exists, therefore. Is this a new stunning quantum-mechanical effect that can only be observed with entangled quantum objects? This is not at all the case! We have analogous experiences in the domain of everyday physics all the time. Here is an example. Somebody owns a pair of gloves and he accidentally takes only one glove with him when leaving the house. So whenever he reaches in his pocket — no matter how far from home he might have departed — and finds a left glove, he knows immediately that the right glove is at home and vice versa. When we, moreover, assume now that he forgets his left or his right glove with the same frequency, we have a situation — however, only as far as the observational results are concerned — as in the case of the two-photon system in state $|\Phi\rangle$ of Eq. (2.7) in the experimental setup of Fig. 2.2.

Is the reader not suspicious yet, he will ask himself at this point at the latest, if not simply classical physics is underlying all the effects of quantum mechanics. Do quantum objects have, in the end, only properties that are classically describable and that can completely explain their behavior? This is at the same time the question whether quantum mechanics can be reduced to classical mechanics. Maybe our efforts for that were not intense enough. Or have we overlooked something? In Section 1.4 we rejected this possibility without any further reasoning. However, using entangled systems, this assumption can be experimentally proven to be wrong.

For an explanation of the measurement results in the Orsay experiment within classical physics, the following assumption could perhaps already be sufficient: a photon has a well-defined property "polarization direction" already before the measurement. This also holds for the single photons of photon pairs. Our source for the photon pairs

---

7) See also the contribution from H. Weinfurter (Chapter 6) in this book. A realization of the entanglement of photons with atoms (quantum interface) is described in the article from G. Rempe (Chapter 5).

is such that always only pairs are produced with either a common polarization direction x or a common polarization direction y. The direction generated in a single process is random but the frequency of the two directions is 50 %, respectively.

What can a proof of the inevitability of quantum mechanics be based on? To bring the measurement process into consideration suggests itself, since we have seen before that regarding this, classical physics and quantum physics are fundamentally different. For this purpose we turn the analyzer in II by an angle Θ, as it is done in Fig. 2.3.

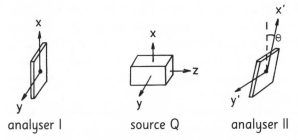

analyser I   source Q   analyser II

Fig. 2.3 Experimental setup as in Fig. 2.2, but with an analyzer in II turned by the angle Θ.

The new directions in II are denoted by x' and y'. When the polarization direction x is measured in analyzer I, the photon II, according to Eq. (2.7), also has the polarization direction x. This is a remarkable consequence of the quantum measurement process, which is independent of the distance between the photons. Finally, according to the rules of quantum mechanics, the polarization directions x' or y' are measured in analyzer II. The respective probabilities are again given by the squares of the projection of the vector $|x\rangle$ onto the vectors $|x'\rangle$ or $|y'\rangle$ (Fig. 2.4). The probability to find the polarization x' is therefore $\cos^2 \Theta$. This is true under the condition that the direction x is found in I, i.e. in half of all cases. For that reason, the probability for the occurrence of the measurement pair (x, x') is in total given by $w(x, x') = 1/2 \cos^2 \Theta$. Depending on which direction is measured in I, four measurement pairs exist and their probabilities $w(x, x')$, $w(x, y')$, $w(y, x')$ and $w(y, y')$ are easily determined as a function of Θ using Fig. 2.4. These probabilities, which are calculated within the framework of quantum theory, are confirmed in fact as relative frequencies in the experiment.

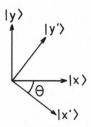

Fig. 2.4 Analyzers turned against each other.

When the reader was thus far believing that everything is classical physics, would he be convinced by this? That is hardly the case, since he will insist that the single photons of the photon pairs are both generated with well-defined polarizations. But now he will allow all possible polarizations for the single photon pairs and give different frequencies for the different combinations. Will he be able to reproduce the experimental results in this way?

Before we answer this question, we will first characterize the point of view of the opponent of quantum theory more accurately: he has no doubts about the measurement results obtained in the Orsay experiment and he is also not questioning the fact that quantum theory provides a correct prognosis of the measurement results. He only insists that this would also be possible within the framework of classical physics. With this, he would like to see the following elements of a classical theory being "maintained" for the phenomena in the quantum domain: first, *realism* (the single quantum objects — also the ones in an compound system — have their properties, which are revealed by the measurement, independent of the measurement, i.e. already before it), and second *locality* (the quantum objects, separate from each other in space, have all their properties independent of each other).[8] A physical theory that complies with this is called a *local realistic theory*.

Our sceptic is convinced that a more detailed description of nature than quantum theory exists, for which all predictions of single measurement results are well defined. The fact that so far only probability statements could be made in the quantum domain is due to the

8) This requirement, which opposes the idea of entanglement of the subsystems, should better be called separability. The entire system disintegrates into subsystems independent of each other. The case where the subsystems are moreover spatially separated from each other is just a special case.

incomplete theoretical description of our quantum objects. A more detailed classical description will be based on additional data. These are just not available up to now, because, for example, appropriate measurement techniques are missing. These hypothetical additional quantities, which according to this conception really exist in nature and allow a local realistic description, are named *hidden variables*. The famous gedanken experiment proposed by Einstein, Podolsky and Rosen (Einstein et al. (1935)) initiated the consideration of hidden variables. The authors were asking the question: "Can a quantum-mechanical description of physical reality be considered complete?", and in this way they were expressing their discontent with the probability structure of quantum theory and the nature of reality depicted by this theory.[9] Rarely in the history of physics has a work describing a conceptional dilemma initiated new developments of that same theory it was trying to question, in such an innovative way.

It is, however, precisely the nature of these hidden variables or parameters according to this concept, that we have no detailed knowledge about them up to now. This opens up a great freedom for the creation of such theories with many possibilities for evading the issue. At first sight, the assertion that classical physics can explain everything appears to be not refutable. But in fact this is possible. Certain measurement results cannot be explained with a local realistic theory — no matter how it is formulated in detail. Quantum theory, on the other hand, is able to explain them. Our opponent of quantum mechanics can thus be proven wrong. This is a fundamental observation that goes back to J. S. Bell. We are going to illustrate it for the experiment described above.

2.5 Bell's inequality: limits of classical physics

In order to prove that local realistic theories are wrong, we — following our strategy further on — have to increase the complexity of our measurement process by considering not just a single rotation of the analyzers I and II against each other but by combining the results from rotations by different angles.

---

9) Further details and the description of the historical situation can be found in the following article from C. Held (Chapter 3).

The Orsay experiment is based on a procedure in which four different orientations in space are used for the analyzers I and II. The corresponding directions x, x', x'', x''' are shown in Fig. 2.5 for an angle Θ. The corresponding perpendicular polarization directions y, y', y'' and y''' are not drawn for reasons of clarity. Using square brackets, we introduce the denomination [analyzer I, analyzer II] for the orientations of the analyzers I and II. It is sufficient to fix the different x-directions. Measurements are carried out for the directions $[x, x']$, $[x', x'']$, $[x'', x''']$ and $[x, x''']$. Four pairs of measurement results can occur in each case. The orientation $[x,x']$ for example can give (x, x'), (x, y'), (y, x') or (y, y').

Fig. 2.5 Different orientations of the x-directions of analyzers turned against each other.

We are going to gain a correlation coefficient from the measurement pairs. For this purpose, we introduce a quantity $s(x)$ for the description of the measurement results. $s(x)$ has the value $+1$ when the polarization x is registered by the analyzer with the orientation x and -1 when the polarization y is recorded. The same holds for $s(x')$, when the polarization x' or y', respectively is measured etc. We can assign to a single double-measurement with the orientation $[x, x']$ the value $+1$ or -1 of the product $s(x)s(x')$ as some kind of a total measurement result, and correspondingly $s(x')s(x'') = \pm 1$ with the orientation $[x', x'']$, etc.

We carry out measurements for many photon pairs and form, for example for the orientation $[x, x']$, the mean value of $s(x)s(x')$:

$$E(x, x') = \overline{s(x)s(x')} \qquad (2.8)$$

The horizontal line marks the averaging. The mean values for the different orientations are finally summed up to the *correlation coefficient S*. It depends on the rotation angle Θ and is defined as follows:

$$S(\Theta) = E(x'', x') + E(x'', x''') + E(x, x') - E(x, x''') \qquad (2.9)$$

Note the change in the sign. Not referring to any theory so far, we are only processing measurement results to a curve $S(\Theta)$. For $\Theta = 22.5°$ this experimental curve shows a maximum: $S_{max}^{Exp} = 2.697 \pm 0.015$.

Is it possible to derive this experimental curve within the framework of a local realistic theory with hidden variables? In such a theory, the single photon pairs leaving the source are described with all the properties of their two photons by a parameter λ. Whatever is measured at analyzer I for one of the photons is independent of the result that is measured in II for the other photon. When the parameter λ is fixed and, for example, the analyzer orientation x (let it be in I or II) is chosen, the corresponding result for the quantity s is fixed. Therefore, a function $s_\lambda(x)$ unknown to us exists, which can have the following values as a function of the parameter λ and the analyzer orientation x:

$$s_\lambda(x) = \begin{cases} +1 \\ -1 \end{cases} \tag{2.10}$$

In this way, the following is represented: the result in analyzer I is independent of the result in II (locality) and the result for the species with parameter value λ is already fixed before the measurement (realism). Nevertheless, correlations can show up for the pairs of measurement results.

We still have to characterize the source regarding the relative frequencies of the different kinds of pairs emitted. For this purpose, we introduce a probability density $\rho(\lambda)$. The probability that a source emits a photon pair with the parameter λ out of the interval $[\lambda, \lambda + d\lambda]$ is given by $\rho(\lambda)\, d\lambda$. It is not necessary to know explicitly both the function $\rho(\lambda)$, which is normalized by $\int \rho(\lambda)\, d\lambda = 1$, and the function $s_\lambda(x)$. Our argumentation is therefore completely general. For a theory with hidden variables (HV), the mean value of E of Eq. (2.8) is formed, for example, for the analyzer orientations $[x, x']$ in accordance with the equation

$$E^{HV}(x, x') = \int s_\lambda(x) s_\lambda(x') \rho(\lambda)\, d\lambda \tag{2.11}$$

With this, also the correlation coefficients $S^{HV}(\Theta)$ in Eq. (2.9) are defined.

The detailed evaluation[10] results in the *Bell's inequality*, which is valid for all angles Θ[11]

$$\left| S^{\mathrm{HV}}(\Theta) \right| \leq 2 \tag{2.12}$$

This is not an equation of quantum theory but — as shown by the derivation — a relation of classical mechanics with great consequences. In view of the previously mentioned experimental result $S^{\mathrm{Exp}}_{\max}$ that is significantly higher than 2, we come to a remarkable conclusion: the measurements violate Bell's inequality (2.12), therefore the predictions of all local realistic theories are false. The attempt to formulate a classical theory for the quantum domain is therefore proven wrong by experiment.[12]

It should be added that starting from the state (2.7) and following the rules of quantum theory, the curve $S^{\mathrm{QM}}(\Theta)$ can also be calculated. We have already taken a first step in this direction with the determination of $w(x, x')$ in the previous section. The result is that correlations may be greater in quantum physics than in classical physics. Once again it is entanglement that makes this possible. $S^{\mathrm{QM}}(\Theta)$ agrees very well with the experimental curve $S^{\mathrm{Exp}}(\Theta)$. Is there anyone who might have expected a different result anyway?

10) The calculation is simple. The term (2.11) is substituted in Eq. (2.9) and the result is suitably arranged:

$$S^{\mathrm{HV}} = \int \left\{ s_\lambda(x'') \left[s_\lambda(x') + s_\lambda(x''') \right] + s_\lambda(x) \left[s_\lambda(x') - s_\lambda(x''') \right] \right\} \rho(\lambda) \, \mathrm{d}\lambda$$

When taking into account that s_λ, according to Eq. (2.10), can only have the values $+1$ or -1 and going through the different combinations, one can easily realize that the absolute of the integrant is always smaller than or equal to 2: $|\{\ldots\}| \leq 2$. With this, and the fact that the probability density is positive ($\rho \geq 0$) follows

$$\begin{aligned}
\left| S^{\mathrm{HV}} \right| &= \left| \int \{\ldots\} \rho(\lambda) \, \mathrm{d}\lambda \right| \\
&\leq \int |\{\ldots\}| \rho(\lambda) \, \mathrm{d}\lambda = \int |\{\ldots\}| \rho(\lambda) \, \mathrm{d}\lambda \leq 2 \int \rho(\lambda) \, \mathrm{d}\lambda = 2
\end{aligned}$$

For the last equation, we have used the fact that the probability density is normalized $\int \rho(\lambda) \, \mathrm{d}\lambda = 1$.

11) Bell's inequality was established by J. S. Bell in 1964 (Bell (1964)). The variant presented here comes from J. F. Clauser, M. A. Horne, A. Shimony and R. A. Holt (Clauser et al. (1969)). Other variants exist. See also the contribution of H. Weinfurter in this book.

12) Of course, the possibility is still open to construct a nonlocal alternative theory to quantum theory. Such a theory with hidden variables is Bohm's theory (Bohm (1952)).

Even our opponent of the quantum theory was not denying that this theory gives a correct prognosis of the experimental results. He was just unsatisfied with it based on his categories for the assessment of theories. Among other problems, he considered quantum theory to be incomplete and he was trying to deduce the measurement results obtained in the quantum domain from a classical alternative theory. This attempt has failed for all local realistic theories because of the generality of the derivation of Eq. (2.12), and with that for a whole class of alternative theories. One could still try it with nonlocal but, nevertheless, realistic theories. But this kind of theory would appear to many even more paradoxical than quantum theory. So we better stay with the well-elaborated and for many decades repeatedly confirmed quantum theory.

However, we stay with all prudence. With the Orsay experiment, we have described a laboratory experiment from the year 1982. Today, the effect of quantum correlations can be demonstrated for distances of more than ten kilometers (Tittel et al. (1998)). Quite soon, weak points of the Orsay experiment were discovered, which showed that the exclusion of local realistic theories is not absolutely compelling. Therefore, more work has been done meanwhile for the refinement of the experiment and other experiments were suggested in order to close the "loopholes". Today one can exclude by the experimental setup that any information about the set polarization is transferred between the detectors — no matter in which way and even faster than the speed of light. Another loophole is due to the fact that detectors do not work with one hundred percent efficiency. Different experimental groups are working on this problem with great success. The next step is probably the use of entangled atoms or ions in ion traps instead of photons. As always, the experimental accuracy for the confirmation of theoretical statements can and must be further improved. However, today the conclusion drawn above can already be regarded as being experimentally extraordinarily well confirmed.

This does not mean that it is impossible to open up completely new domains of experience with new experimental techniques and observation methods, for which a quantum-theoretical description of the present kind fails. Should this succeed one day, then a domain outside the quantum domain would be found, which will require a new theory. Nobody, however, expects that this could be a local realistic theory. In the meantime, we will try to extend our understanding of quantum theory. For this, we proceed to a new example.

2.6 The eavesdropper is detected: quantum cryptography

With the experimental setup described in the previous chapter, we can directly demonstrate how entangled systems can be used to transfer secret messages so that an "eavesdropper" has absolutely no possibility to "tap" them (Ekert (1991)). The quantum theory opens up completely new possibilities for *cryptography*. And again this is essentially based on the fact that in quantum physics a measurement — different from classical physics — in general changes the state of an object. Classically stored information allows reading or listening free of interference. When, in contrast, someone tries to "tap" by measurement the information stored in a quantum state, the quantum system is transferred to a new state. The eavesdropper leaves a trace behind. The "bugging" can be uncovered.

When an eavesdropper, for example, tries to measure the state of polarization of photon 1 or photon 2 on the way to I or II, he will, in general, change its polarization even if the corresponding photon is not destroyed by this. After this measuring invasion, the composite system is no longer in the entangled state (2.7), but in a product state. When the above-described measurement for the determination of $S(\Theta)$ is made in I and II afterwards, the quantum-theoretical maximum value $S_{\max}^{\mathrm{QM}}$ is no longer found. The correlations are changed and the eavesdropper is uncovered. The use of the two-photon system from the previous section has yet another advantage. We will see that with this system, a completely randomly-formed key can be transferred from the photon source to I and II. How can these elements be combined for a secure coding system?

At first we will describe one of the possible *coding procedures*. We assume that the text is written in bits. It thus consists of a sequence of the numbers 0 and 1. The key is also a sequence of numbers 0 and 1. It is added modulo 2 to the text that is to be encoded (hence by using the rule $0 + 0 = 0$, $0 + 1 = 1$, $1 + 1 = 0$):

	0	1	1	0	0	0	1	text
+	1	0	1	1	0	1	0	key
	1	1	0	1	0	1	1	encoded text

For the decoding, the key is added to the encoded text following the same rules and the original text emerges again:

	1	1	0	1	0	1	1	encoded text
+	1	0	1	1	0	1	0	key
	0	1	1	0	0	0	1	original text

In cryptography the encoded text is transferred over a classical channel (telephone) that can be tapped, or even over public media (newspaper). The key, which should only be available to the sender and the receiver, must on the other hand remain absolutely secret. It is supposed to be a completely random sequence of the numbers 0 and 1 and it must only be used once. The problem therefore is the transmission of the key. The key can either be exchanged between sender and receiver or copies of the key are transmitted to both sender and receiver by a source. We will consider the second case.

The essential point is the following: In order to ensure secrecy, it is sufficient that both sender and receiver can find out whether the key transfer was tapped. After they have convinced themselves that this is not the case, they will use the key. If it was tapped, they discard the key and wait for the source to transmit a new key, which both of them check separately again whether it was tapped, etc. So two procedures are necessary, a key transmission and a bugging test. The entangled photons of the previous paragraph allow for both.

For the *transmission of the key*, the orientation $[x, x]$ of the parallel-oriented analyzers I and II is defined in public, which means the x-direction is fixed in public. We have seen that in this case either the measurement pair (x, x) or (y, y) is registered in a random way. (x, x) is then interpreted in I and II as the transmission of the key number 0. (y, y) is regarded as the transmission of the key number 1. Both numbers appear totally randomly and on average with equal frequency. Now the question is, whether the transmission remained secret. For the *bugging test*, we recall once again the four orientations of the analyzers turned against each other in space in the way they are pictured in Fig. 2.5 (the respective y-directions, not drawn, are perpendicular). We choose the setting $\Theta = 22.5°$. In this case, the quantum-theoretical maximum value S_{max}^{QM} is reached for $S(\Theta)$ in the experiment. When the state of the quantum system, on the other hand, was changed by the influence of the measurement, $S(\Theta)$ is smaller. The key transmission and bugging test are now combined, using in addition a public channel. The observers in I and II, who want to receive a key that is not tapped, are thereby now proceeding step by step:

At first, the four spatial orientations x, x', x'', x''' from Fig. 2.5 are defined in public uniformly for I and II. With this, the corresponding y-orientations of the analyzers are also fixed. The source produces single pairs of entangled photons.

Now under cover and independent of each other, the analyzers are arbitrarily set in I and II for each photon pair anew in one of the four orientations. The orientation of the corresponding analyzer is noted at both places. It is also recorded, if for this orientation the result of the polarization measurement was the x- or the y-direction. This procedure is repeated for many photon pairs.

After all these covered measurements, the public channel comes into action again. The observers in I and II exchange which orientations they had set for the measurement at the first pair, the second pair, etc.

Then each of them sorts out independently under cover and by keeping the temporal sequence for the measurements, if the analyzers in I and II were parallel or turned against each other. With this, we get to the key transmission and the bugging test as follows:

Both observers know when the analyzers were turned against each other. The measurements obtained in these cases are exchanged over the public channel in the next step. Now both of them have all the information that is necessary to calculate $S(\Theta)$ according to our considerations in Section 2.5. Consequently, they can check for themselves if the value of S_{max}^{QM} is realized. When this is not the case, an eavesdropper was trying to tap. The whole series of measurements is discarded and the procedure started over again. In the other case, both observers know that there was no bugging.

Both observers now finally turn to the measurement results obtained for parallel-oriented analyzers, which they kept in secret. We have seen before that the parallel alignment allows the transmission of a key. This transmission has succeeded in principle with absolute secrecy.

Classical cryptography uses elaborated mathematical coding techniques in order to prevent attempts at bugging. These, on the other hand, can be surmounted with mathematical techniques. Someone with a faster computer or with better software can break the code. In contrast, the information in *quantum cryptography* is protected with the laws of physics. They — at least in principle — cannot be circumvented. In which way quantum cryptography can be technically realized is a different question.[13]

2.7 Sheep can be cloned, photons not

As shown before, only probability statements about the result of a single quantum measurement are possible in general. Only after repeating the measurement in principle an infinite number of times by preparing the quantum system always in the same state and carrying out the same measurement, is a particular measurement value obtained with the predicted probability. For example, for the measurement value 1 of the state (1.8), the relative frequency P_1 of Eq. (1.9) results. This suggests that information about the factors c_1 and c_2 and thus about the state $|\Psi\rangle$ of Eq. (1.8) is obtained only when this state is prepared at least very often in the same way and then the procedure according to the description from above is followed. What is known about the determination of a quantum state? Once again we will cast some light on the process of a quantum measurement from a different perspective. And again we are going to encounter entangled systems.

But at first we want to convince ourselves that in the quantum domain — different from classical physics — it is not always possible to distinguish between two arbitrary states by means of one single measurement. An example should clarify this. Assuming that we try to distinguish by one measurement if a system is in the state $|1\rangle$ or

---

13) Experiments for quantum cryptography are described in this book in the contributions of H. Weinfurter. Single photons are needed for the information transfer, so that the entanglement of photons with atoms in different places can be used. For single-photon sources see the chapters by G. Rempe (Chapter 5) and H. Weinfurter (Chapter 6).

the state $\left(|1\rangle + |2\rangle\right)/\sqrt{2}$. When the state $|1\rangle$ is present, we get the measurement result 1 for sure. When the state $\left(|1\rangle + |2\rangle\right)/\sqrt{2}$ is present, the measurement result 1 or 2 is obtained with a probability of $1/2$ in each case. However, we can only measure once. When we get the result 2, we can definitely conclude about the presence of $\left(|1\rangle + |2\rangle\right)/\sqrt{2}$; the result 1 instead gives no useful information at all.

The general theorem behind this observation is: it is in general impossible to distinguish between two nonorthogonal states with one single quantum measurement. In classical physics different states of a system can be distinguished with one measurement, but in the quantum domain this is impossible. This is obviously a very restrictive theorem. But is there any way to get around it easily for all practical applications? For this it would just be necessary to produce many copies of the unknown state and carry out measurements with these copies. For the example above either the measurement result 1 alone or with the same frequency the results 1 and 2 would occur; in this way, the states could be clearly distinguished. The question, however, is whether it is somehow possible to "clone" or copy unknown quantum states.

In fact, the linearity of quantum mechanics, which we learned about in Section 1.6, upsets our plans. Let us assume one could make a copy. The quantum system to be copied, the *quantum copy machine* and the copy itself form a compound system. The quantum copy machine might initially be in the state $|k_0\rangle$. While copying the state $|1\rangle$, it changes itself into the state $|k_1\rangle$; when copying $|2\rangle$, it changes into the state $|k_2\rangle$. Therefore, one of the two following quantum-mechanical processes is going to happen, of which we just write down the initial and the final states:

$$|1\rangle\,|k_0\rangle \rightarrow |1\rangle\,|1\rangle\,|k_1\rangle \tag{2.13}$$

$$|2\rangle\,|k_0\rangle \rightarrow |2\rangle\,|2\rangle\,|k_2\rangle \tag{2.14}$$

The quantum copy machine transfers initial states into final states within the framework of dynamics I. This is made in a linear way (see Eq. (1.13)): when the initial state is a sum of states, the final state results as a sum over the transformed states. In our case, applying the product rule, we get

$$\left(c_1\,|1\rangle + c_2\,|2\rangle\right)|k_0\rangle \rightarrow c_1\,|1\rangle\,|1\rangle\,|k_1\rangle + c_2\,|2\rangle\,|2\rangle\,|k_2\rangle \tag{2.15}$$

Remarkably, an entangled state is thus formed again. The result, however, would have looked different when our machine had really functioned like a copy machine:

$$(c_1 |1\rangle + c_2 |2\rangle) |k_0\rangle \rightarrow (c_1 |1\rangle + c_2 |2\rangle) (c_1 |1\rangle + c_2 |2\rangle) \cdot |k'\rangle \tag{2.16}$$

Here, $|k'\rangle$ is the final state of the copy machine. The result, therefore, should have been a product state and not an entangled state. Only in the case when either c_1 or c_2 equals zero, does copying take place. The consequence is: no quantum copy machine exists that could copy just any state. Quantum states cannot be cloned (Wootters and Zurek (1982), Dieks (1982)). This is also called the quantum-mechanical *no-cloning theorem*.[14] The cloning would violate the linearity of quantum mechanics. With this, our idea for the measurement of an unknown quantum state has failed. The no-cloning attribute appears to be destructive. However, we are going to show that it can also have very constructive consequences. It protects quantum mechanics from a contradiction with special relativity by preventing certain processes.

We return once again to our entangled photons of Section 2.4. Assuming that cloning is possible and having a copy machine for unknown states, we could install it into the experimental process with disastrous consequences.

In order to clarify this, we are first going to add a short calculation. We consider analyzer orientations in the experimental setup of Section 2.4, which are turned by $\Theta = 45°$ against each other. One glance at Fig. 2.4 then shows that $|x\rangle = (|x'\rangle + |y'\rangle) / \sqrt{2}$ and $|y\rangle = (-|x'\rangle + |y'\rangle) / \sqrt{2}$ holds for the state vectors. Inserting this into Eq. (2.7) shows after a short calculation that the state $|\Phi\rangle$ can also be written in the form

$$|\Phi\rangle = \frac{1}{\sqrt{2}} (|x'_1, x'_2\rangle + |y'_1, y'_2\rangle) \tag{2.17}$$

When the analyzer in I is now oriented towards the directions x and y, the overall state, after the measurement in I, is transferred with the probability 1/2 to $|x_1, x_2\rangle$ and with the probability 1/2 to $|y_1, y_2\rangle$ (cf. Eq. (2.7)). After this measurement, the photon running towards the analyzer II got the state $|x_2\rangle$ or $|y_2\rangle$, correspondingly. Equation (2.16)

---

14) The relevance of this theorem for the quantum information theory is addressed by R. F. Werner (Section 7.2) in this book.

shows that the analog holds for the same state $|\Phi\rangle$, when the analyzer in I was turned by $\Theta = 45°$. A measurement is then transferring into the state $|x_1', x_2'\rangle$ or $|y_1', y_2'\rangle$ and immediately the state $|x_2'\rangle$ or $|y_2'\rangle$ is present at the analyzer II. If any kind of possibility to distinguish between $|x_2\rangle$, $|y_2\rangle$, $|x_2'\rangle$, $|y_2'\rangle$ with one measurement existed, a message could be transmitted from I to II using entangled photon pairs: for the transmission of a "0" in I, the orientation $\Theta = 0°$ would be chosen and for the transmission of a "1", the orientation $\Theta = 45°$. This information could immediately be read in II when the photon source is placed in the middle. Obviously in this case special relativity would be violated. However, since we have seen at the beginning that it is not possible to discriminate between the prime and the not prime state by one measurement, the information can actually not be transmitted and the special relativity theory appears to be saved.

But in fact this is not totally the case yet. If it were possible to install a quantum copy machine in II, the state arriving there could be copied quickly many times and measurements could be carried out with analyzers in xy- and $x'y'$-orientations. Then the state present in II could be determined and a signal transmission faster than the speed of light would have succeeded. Only the nonexistence of a quantum copy machine prevents this possibility as well.

Therefore, one cannot succeed to establish a contradiction between quantum theory and special relativity theory by using entangled quantum systems. Quantum theory protects itself from that. This is once again a really amazing result, because quantum theory and relativity theory are two theories with completely different physical concepts. They have been formed totally independent of each other. Under this aspect, Einstein actually should have got to like the quantum theory, what he declined to do.

2.8 The parts and the whole

In conclusion, we want to take an interpreting glance backwards. For a compound quantum system in an entangled state, one can still speak about the single systems. These can be, for example, two photons flying in different directions at which polarization measurements are carried out. The measurement can take place several kilometers distant from each other. Even though there is no interaction at all between the subsystems, the systems turn out to be — in a non-classical way — correlated. The quantum world is *nonlocal*.

The example from Eq. (2.4) gives evidence that in the case of entanglement the subsystems cannot be assigned with their own state vector, in contrast to the product states in Eq. (2.3). As we already know, a state vector can in general be ascribed to single systems but not always properties. For example, only after the measurement of "position" do quantum systems also have the property "position" and this property has a specific value. In the case of entangled systems, we have seen that for the subsystems the equivalent statement is true for the state itself. Only after the measurement at a subsystem can a state in general, be ascribed. In an entangled compound system, the subsystem does not even have the attribute of having its own state anylonger.[15] The list of the things that got lost in quantum physics compared to classical physics has to be extended further with the "state of a subsystem" in an entangled compound system.

The wording "got lost" should not be understood in a negative way. This is rather a positive characterization. Quantum theory is more general than the classical theories. Quantum theory goes beyond the framework that limits these theories. However, one problem immediately suggests itself: How can classical physics be regained as a limiting case of quantum physics? An answer to this question does not exist up to now, but there are approaches for a solution.[16] Obviously the problem was aggravated by the possibility of entanglement: Why can entanglement be established for example for two photons, but never for two chairs or other classical objects? We have said above that entanglement would be the "normal case". This statement must be restricted to the quantum world, but what are the theoretical reasons for this?

In classical physics, the states of the subsystems determine the state of the entire system (*separability*). When it is impossible to separate a compound system, the system is called *holistic*. With the entangled quantum systems, a true holism was found in nature.[17] In the case of

---

15) We are speaking about pure states represented by vectors as described in Section 1.6. Subsystems may still be described by density operators.
16) See the article of E. Joos (Chapter 8) in this book.
17) The consequences for philosophy of nature are discussed in the chapter by M. Esfeld (Chapter 10).

entangled systems, the whole is more than the sum of its parts even though there is no interaction between the parts.[18]

References

- A. Aspect, P. Grangier, G. Roger (1981), *Phys. Rev. Lett.* **47**, 460.
- A. Aspect, P. Grangier, G. Roger (1982), *Phys. Rev. Lett.* **49**, 91.
- J. Audretsch, K. Mainzer (1990 and 1996) (eds.), „Wieviele Leben hat Schrödingers Katze? Zur Physik and Philosophie der Quantenmechanik", Spektrum, Akad. Verlag, Heidelberg.
- J. Audretsch (2005), „Verschränkte Systeme — die Quantenphysik auf neuen Wegen", Wiley-VCH, Weinheim. An English translation of the textbook with the title "Entangled Systems" will be published in 2006 by Wiley-VCH, Weinheim.
- I. G. Barbour (1966), "Issues in Science and Religion", ICM Pr., London.
- J. S. Bell (1964), *Physics* **1**, 195.
- D. Bohm (1952), *Phys. Rev.* **85**, 166, 180.
- J. F. Clauser, M. A. Horne, A. Shimony, R. A. Holt (1969), *Phys. Rev. Lett.* **23**, 880.
- D. Dieks (1982), *Phys. Lett.* **A92**, 271.
- A. Einstein, B. Podolsky, N. Rosen (1935), *Phys. Rev.* **47**, 777.
- A. K. Ekert (1991), *Phys. Rev. Lett.* **67**, 667.
- P. G. Kwiat, A. M. Steinberg, R. Y. Chiao (1992), *Phys. Rev.* **A45**, 7729.
- R. J. Russel (1988), "Quantum Physics in Philosophical and Theological Perspective", in: "Physics, Philosophy and Theology: A Common Quest for Understanding", R. J. Russel, W. R. Stoeger, G. V. Coyne (eds.), p. 343–374, Vatikan city.
- E. Schrödinger (1935), *Naturwiss.* **23**, 807–812, 823–828, 844–849.
- M. O. Scully, B.-G. Englert, H. Walther (1991), *Nature* **351**, 111.
- W. Tittel, J. Brendel, B. Gisin, T. Herzog, H. Zbinden, N. Gisin (1998), *Phys. Rev.* **A57**, 3229.
- W. K. Wootters, W. Zurek (1982), *Nature* **299**, 802.

---

18) It is scarcely known that the consequences of quantum theory for philosophy of nature received attention right into theology (see among others Russel (1988) and Barbour (1966)).

3

The Bohr–Einstein debate and the fundamental problem of quantum mechanics

Carsten Held

At the beginning of the 20th century, two developments in physics revolutionized our scientific picture of the world: relativity theory and quantum mechanics. Albert Einstein boldly initiated the first of these two revolutions by developing the theory of relativity; our ideas of the connection of space, time, and matter were fundamentally changed by this theory. In particular, the physics of the macroscopic domain, the universe as a whole not only verified Einstein's theory, but today is inconceivable without it. Einstein, however, was from the beginning sceptical of quantum mechanics — a new theory of the microscopic domain, of atoms and elementary particles — and finally even disapproving. This theory had reached its preliminary closed form in the year 1927 and Niels Bohr, one of its fathers, attempted a first interpretation in a lecture at the Solvay conference in Brussels; the complementarity interpretation was to show clearly that quantum mechanics is not a patchwork but a comprehensive and fundamental theory of matter. In the discussion following Bohr's lecture, Einstein publicly opposed him: quantum mechanics, he said, might at best be a transition state on our way to finally understand the micro-objects and their interactions but is surely no final fundamental theory. The "inevitable chance" built into quantum mechanics cannot be the last word on the structure of matter, because this would mean an inevitable limitation for our effort to understand and explain this structure. If we imagine the world being made by a creator, this was probably done in a way that he in principle allows us to under-

Entangled World, Jürgen Audretsch
Copyright © 2005 WILEY-VCH Verlag GmbH & Co. KGaA, Weinheim
ISBN: 3-527-40470-8

stand, because — according to a famous saying by Einstein — "subtle is the Lord but vicious he is not".

Still on the conference in Brussels, Einstein improvized gedanken experiments, which were supposed to show how the natural limitations of quantum mechanics could be outwitted, in order to learn more about the micro-objects than what they divulge to us by themselves. Bohr, however, succeeded in discussions, which in retrospect he called "dramatic", to refute all these arguments. This was the beginning of a dispute that lasted for decades: the Bohr–Einstein debate. In the following, this debate will be presented in broad outline. Commonly, the view is taken that Bohr, the advocate of the new quantum mechanics, has gained the victory in the dispute with Einstein, the critic of the theory, but this is not entirely correct. Surely in the first part of the discussion, in which Einstein still tried to circumvent Heisenberg's uncertainty relations, Bohr was able to refute each of his gedanken experiments. In the discussions after 1930, however, Einstein changed his tactics. Now he accepted the uncertainty relations and argued in summary as follows: even though according to the relation certain properties of a quantum object cannot, out of principle, be measured exactly at the same time, they must be instantiated in the object; quantum mechanics cannot be a complete description of physical reality. The result of this new tactics was the EPR article published together with the young physicists Podolsky and Rosen in 1935. The authors designed a conflict between the assumption that quantum mechanics is complete and the assumption that only local causal interactions exist between physical systems. Bohr's reply to this argument is so ambiguous that it cannot be taken as a part of an understandable interpretation of the facts (cf. Bell (1987), 155f.). Nevertheless, this reply contains a core that contributes to accurately bringing out the fundamental problem of understanding quantum mechanics. The goal of this chapter is to show that Bohr and Einstein were moving towards a single problem in their dispute, which in fact constitutes the fundamental problem for the understanding of quantum mechanics.

3.1 Einstein's argument against quantum mechanics as a complete description (1927)

Paul Ehrenfest had already written about himself, Einstein and Bohr in a letter to Einstein in 1925:

> I know that no living human being has glanced so deeply into the essential abysses of quantum theory as you two, and that nobody but you really sees how entirely radically new conceptions are necessary.[1]

By that time Ehrenfest could not yet know that Bohr was going to be the one to develop "radically new conceptions", and Einstein, in contrast, the one to radically reject these conceptions and require that quantum mechanics be made intelligible on the basis of classical scientific rationality. At the mentioned Solvay conference in Brussels two years later (1927) Ehrenfest was present, too. Here, he experienced first-hand the dramatic beginning of the Bohr–Einstein debate, the personal dispute between the two physicists, their first wrestling with the "radically new conceptions". One part of this struggle took place on the scene, in the public arena of the conference, the other behind the scenes.

The public part of the discussion went as follows: Bohr had given a lecture on his new interpretation of quantum theory.[2] There, he discussed right at the beginning intricate epistemological questions, in order to introduce his new fundamental concept of complementarity; as we know today, he had labored most intensively on this passage before the conference. This clearly shows his effort to especially convince Einstein of the new concepts of quantum mechanics. However, it also shows clearly that Bohr himself was above all trying to solve one problem: the wave–particle duality. How is it possible that certain subatomic objects, e. g. photons or electrons, appear to us as waves on one occasion, as particles on the other? This problem — a fictitious problem, as it will turn out — was going to be solved by the concept of complementarity.[3]

The meaning of the expression complementarity could not become clear to Bohr's listeners at once, because the new conception was too

---

1) Ehrenfest to Einstein, 16. 9. 1925 (cf. Held (1998), 11 f.).
2) The German print version of this lecture is the article "Das Quantenpostulat und die neuere Entwicklung der Atomistik" (Bohr (1931), 34–59).
3) Compare to Section 5.1.

difficult, even paradoxical. The meaning is approximately the following: two descriptions of an atomic object are complementary, if they on the one hand exclude each other, but on the other hand are both necessary for a complete description of the object. However the new conception is fundamentally incomprehensible. This becomes clear in the way it is used by Bohr on the one hand for the classical images of physical objects — wave-image and particle-image — on the other hand for certain quantum-mechanical measurement quantities, e. g. position and momentum, and that it has a completely different meaning in each case. Complementarity does not explain the relation of these images or quantities, but instead is explained by them, and in fact in two entirely different ways. The wave-image and the particle-image exclude each other in classical physics, while in quantum mechanics they supposedly complement each other. The values of quantities like position and momentum, however, complement each other in classical physics, but are supposed to exclude each other in quantum mechanics. And even worse: for the wave- and particle-image complementarity is supposed to show that both together are correct — the classical exclusion and the quantum-mechanical complementation. For the values of position and momentum, however, complementarity is supposed to show that only one is correct — quantum-mechanical exclusion. A confusion that can hardly be outdone! When we now in addition get to know that the mutual exclusion of wave-image and particle-image is not a question of classical physics but that the images, understood as dynamical descriptions of physical objects, exclude each other logically, while according to Bohr they are also supposed to complement each other for an integral description of an object (and this without contradiction), then finally the suspicion arises that the complementarity of wave and particle is a nonconcept that cannot be reconstructed, an attempt to solve a paradox by means of a paradox.

Moreover, in the case of the values of quantum-mechanical quantities like position and momentum, complementarity does not fare any better. These values do not exclude each other at all, neither logically nor classically, nor quantum-mechanically. Only their simultaneous *determination* at a quantum object is excluded due to Heisenberg's uncertainty relation for position and momentum. Now, we know that according to Bohr these values do exclude each other in the sense that they cannot simultaneously *exist* at a quantum object. But what then is the purpose of the complementing aspect of complementarity? A comprehensible answer can be gleaned only from Bohr's articles from 1935 on. The concept of complementarity here refers only to the values of quantum-mechanical quantities for which the uncertainty relation holds. This means that these properties are finally just complementary in the sense that they once have complemented each other in classical physics, but exclude each other according to the results of quantum mechanics.[4] All this, however, was not clear to Bohr in 1927 when the concept of complementarity was coined; it became clear to him through the years of discussions with Einstein.

In October 1927 at the Solvay conference in Brussels Einstein had the opportunity in a plenary discussion to respond directly to Bohr's lecture. However, he did not say a word about it; he addressed neither the new concept of complementarity, nor the problem of wave and particle. Instead, he asked himself and the public the question: Can the new quantum theory be regarded as "a complete description of the individual atomic processes"? For this purpose he exemplified a little gedanken experiment, which had nothing to do with Bohr's lecture. Bohr for his part might have reacted to this contribution, however, he did so just as little — and he can hardly be blamed for that. Instead he tried to re-interpret Einstein's example (as far as he had understood it at that point) as a wave–particle problem and a case of complementarity. In short: in their first public confrontation, the two physicists completely talked at cross-purposes.

In his contribution to the discussion, Einstein considers a diaphragm with a slit aperture, through which electrons are passing. A suitable screen is placed behind, which registers the electrons. The electrons are described as waves that penetrate the slit and reach the screen as spherical waves. Their intensity on the screen is a measure for "whatever happens at this position". This is the idea of Born's probability interpretation: the single wave (ψ) determines the probability ($|\psi|^2$) to find the electrons at certain positions.[5] Now Einstein sets two interpretations of this wave in contrast, namely on the one hand that a single wave "does not correspond to a single electron but to an electron cloud", and on the other hand that this is exactly what it does, that is each single wave corresponds to a

4) In the following, the expression "complementarity" should be understood only in this sense.
5) Compare to Section 1.6.

single electron. One gets to the last interpretation starting from the following basis: the theory is a complete theory of the individual process. According to Einstein, this is supported by the fact that only in this way can important experiments be explained by quantum mechanics.[6] But now Einstein gives an objection, which is obviously more important from his point of view:

> The scattered wave [ψ] that moves towards P [the screen] does not present any preferred direction. If $|\psi|^2$ were simply considered as the probability that a definite particle is situated at a certain place at a definite instant, it might happen that *one and the same* elementary process would act at two or more places of the screen. But the interpretation according to which $|\psi|^2$ expresses the probability that *this* particle is situated at a certain place presupposes a very particular mechanism of action at a distance, which would prevent the wave continuously distributed in space from acting at *two* places of the screen. (Bohr, *Collected Works* **6**, p. 101 f.)[7]

Here, Einstein points to a fundamental difficulty for the interpretation of quantum mechanics: the idea that the theory is a complete description of physical systems leads to an implausible, even mysterious long-distance effect, the so-called collapse of the wave function. Einstein's argumentation should be examined in more detail. If quantum mechanics describes the single processes completely, obviously the wave describes a single electron completely. Then, however, this electron behind the slit must be distributed or spread over a semicircle for the following reason: the diffracted wave — according to Einstein — "shows no preferred direction", which means that in a semicircle no direction is distinguished from all others. However, when nothing supposedly really exists besides what is given by this function, also the electron does not move in a certain direction but spreads out uniformly in all directions. The wavefunction describes the motion of the electron as a wave motion and this can obviously have only one meaning: the electron *is* a wave, at least in the sense that it spreads out like a wave. This, of course, is still the case when the wave hits

6) At this point Einstein strangely thinks of the Geiger–Bothe experiment (cf. Bohr, *Collected Works* **6**, 102) that confirms the photon picture of light (cf., e. g., Murdoch 1987, 27–28).
7) My italics.

the screen and covers it entirely: the electron itself is what covers the screen as a whole.

The assumption of a wave-nature of the electron does not appear to be completely implausible at the time when the wave- and particle-nature of matter were so little clarified, and Einstein with no syllable gives the impression that he would regard this idea as absurd. Now the wavefunction according to the precondition also specifies the probability for the electron to hit on certain positions of the screen. These positions are regarded as being very small regions, ideally points on the screen, in sharp contrast to the wave ψ itself, which covers the whole screen. So the probability interpretation of the wavefunction presupposes that the electron can hit at *certain positions*. This, however, can only mean that the electron is a particle when it hits, at least concerning its localization at a certain position. The "completeness interpretation" says even more exactly: the wave function determines for a single particle the probability[8] to hit certain positions of the screen. Therefore, it must at least be possible that the electron hits just any, but a *definite* place. This trivial consequence of the probability interpretation of the wave function only reflects what is really seen in such an experiment: the electrons indeed hit certain positions of the screen.

Thus the single electron on the one hand is located at a certain position when it reaches the screen — this results from the probability interpretation and is, as we said, also observed. On the other hand — as clarified above — the single electron should not hit a certain position, but be equally distributed over the whole screen. Obviously the two interpretations contradict each other. How could the contradiction be avoided, how can the appearance at one place on the screen and the distribution over the whole screen be brought together? At this point, Einstein introduces a "special action-at-a-distance mechanism". This means the following: the wave has an effect on the whole screen, but it is sufficient to regard two arbitrary places A and B where it acts (Einstein says the wave acts "at *two or more* places on the screen"). Now the assumption from the probability interpretation is added — one electron hits at a definite position on the screen — and according to the completeness interpretation exactly one electron is described by the wavefunction, so that this can be expressed as the impact of

---

8) Compare to Section 1.6.

exactly one electron. When this electron is now measured, e. g., in A, it exists in A and not as assumed before, in A and B. To prohibit any simultaneous effect of the wave at point B, a breakdown of the wave in the course of the measurement from the whole width of the screen to the place A must occur, a process that was later termed as "collapse of the wavefunction". Einstein emphasizes the implausible aspect of this process by pointing to a "contradiction with the postulate of relativity". No matter how distant A and B are from each other (in the language of relativity theory: whether they are space-like separated or not), the collapse in B must happen exactly simultaneously with the measurement in A. In short: Born's probability interpretation and the idea that quantum mechanics completely describes single processes led to a conflict with the theory of relativity.

However, Einstein's aim in his criticism is not to show a logical conflict of two physical theories, quantum mechanics and relativity theory — for example in such a way that one theory forbids propagations of effects faster than the speed of light, while the other theory requires them. His intention is rather to show that a certain *interpretation* of quantum mechanics, which regards this theory as being a complete description of single micro-objects is absurd, because it leads to a physically implausible "collapse" of a wave to one point. The theory of relativity is only used to exhibit the implausibility of this interpretation by an explicit contradiction with an accepted theory. Einstein says, after having specified the problem, that in order to solve the contradiction one has to assume a "very special action-at-a-distance mechanism" and he continues:

> In my opinion one can only counter this objection in the way that one does not only describe the process by the Schrödinger wave, but at the same time one localizes the particle during the propagation. [...] If one works exclusively with the Schrödinger waves, interpretation II of $|\psi|^2$ in my opinion implies a contradiction with the relativity postulate. (Bohr, *Collected Works* **6**, p. 101 f.)

This is to say, with the "special action-at-a-distance mechanism" itself the essential "objection" has already been made. Einstein may expect his listeners to regard this obscure mechanism itself as absurd, or to recognize the additional "contradiction to the postulate of relativity" — but it is this action-at-a-distance mechanism, and thus the "collapse" that is problematic by itself.

In this way, Einstein illustrates the whole dilemma of the interpretation of quantum mechanics. Assuming that the quantum-mechanical description — here represented by a semic-circular wave in space — were a complete description of the single electron, the electron at the time of impact would be spread out over the whole screen and not localized in one place. A mediating process has to be taken into consideration due to the fact that it can be proved that the electron hits at a well-defined position — the "collapse of the wave function". Since this process in principle has to take place in an infinitely short time and distributed over a very large space, it appears to be physically implausible. It seems much more plausible to assume that the "process is not only described by a [... wave], but at the same time the particle is localized during the propagation". The particle actually needs to be described in a way that goes beyond the wave. The quantum-mechanical description (as a wave) is thus *not* complete. There is still a hidden quantity, the unknown "localization" of the particle, which actually determines its behavior.

This is the fundamental problem: either quantum mechanics is incomplete and needs to be completed by a theory of hidden quantities, or it is complete and then the collapse of the wavefunction must be made physically plausible. This dilemma has not been solved until today, but on the contrary has become more and more critical. Nowadays, we possess strong mathematical arguments, showing that it is impossible to complement quantum mechanics by theories of hidden variables under certain completely plausible conditions. Therefore, the problem of the collapse remains. The collapse is an especially radical effort to solve the so-called measurement problem, the central and still controversial interpretation problem of quantum mechanics.[9] In

9) The measurement problem is in its most general form the problem of making the two following statements consistent with each other: (1) "The system exists in a superposition of eigenstates of the observable A" and (2) "the property a_n of A is measured". If one tries to solve this problem by describing besides the measured system also the measuring apparatus in terms of quantum mechanics (which must be possible in principle), the result is that also the apparatus ends up being in a superposition state. This means that, under a certain additional condition, no value is shown — in contradiction with the direct perception of such values in apparatuses upon measurement. Often in the literature, only this derived case is regarded as the measurement problem. The derivative problem is of course not solved by the assumption of a collapse in the beginning of the interaction with the measuring apparatus, but avoided right from the beginning.

a theory of hidden quantities[10], this problem would not emerge at all.

3.2 Einstein's attempt to avoid the uncertainty relations and Bohr's counterarguments (1927–1930)

The first public confrontation between Bohr and Einstein took a superficial course, as we have seen. Behind the scenes, however, the dispute was continued in concentrated manner. Paul Ehrenfest was also there as an eye and ear witness. He describes the intensive disputes in a letter:

> "It was enthralling for me to join the dialogues between Bohr and Einstein; like a game of chess. Einstein [invented] new examples again and again. In a way a perpetuum mobile of second order, to break through the *uncertainty principle*. Bohr always picked his tools out of a dark cloud of philosophical smoke, in order to break example after example. Einstein like a jack-in-the-box: jumped out refreshed each morning. Oh, it was exquisite. But I am almost without reservation pro Bohr and contra Einstein. He behaves now against Bohr exactly as the defenders of absolute simultaneity behaved towards him."[11]

Ehrenfest's report indicates what these discussions were about. Einstein devised a series of gedanken experiments, meant to avoid the "inaccuracy relations" as Ehrenfest used to call them. He means the Heisenberg relation for position and momentum mentioned above. This relation says that the position and momentum of a quantum object cannot be measured simultaneously beyond a certain limit of accuracy. Einstein understood this relation only as a limitation of what can be measured directly at a single atomic object. Now he wanted to indirectly obtain more information from these objects than is possible according to the uncertainty relations. For this purpose he considers, in addition to the quantum object he is interested in, a sec-

10) Compare to Section 2.4.
11) Ehrenfest to Goudsmit, Uhlenbeck and Dieke, 3. 11. 1927 (in German in Bohr, Collected Works 6, 415 f.).

ond object, which has been in interaction with the first one before and can now give information about it. In this way he wanted to show that the quantum-mechanical description is incomplete — just by showing that certain properties of quantum objects do not exist in this description but can be measured indirectly via the second object.

Among other things, Einstein analyzed once again his gedanken experiment described above: a particle beam hits a slit with a screen behind. This time the diaphragm is equipped with a freely mobile slide, which can be moved by a passing electron. In this way one could determine the direction in which the particle flew away and thus (while apart from this the experiment is running as before) the path of the particle between diaphragm and screen could be predicted more accurately than is allowed by the half-circular wave function. We can clearly see from this proposal that Einstein that he was not assuming that the electron behind the aperture would indeed spread half-circularly, i.e. that it *is* a wave. The electron has to take a certain path, which the description by a wave does not reflect; this path can possibly be detected indirectly using the trick of the mobile diaphragm.

Bohr brought his counterarguments: Einstein's conclusion that the direction into which the electron moves behind the diaphragm could be determined more accurately than the wave function permits, while the experiment except for this could run *exactly* like before is wrong. Bohr underlined: the mobile slider is a macroscopic object for which the uncertainty relation in general can be neglected. In the gedanken experiment, however, the mobile slider becomes a second quantum object because it has to interact with the particle to give us information about it. The situation we have here is therefore nothing but a quantum-mechanical two-body problem, comparable, e.g., to the Compton effect (cf. Bohr (1949), p. 127 f.).

This reply, reconstructed from Bohr's own later notes is at best the sketch of an argument, which can be reconstructed as follows. The description of the experiment using a certain wavefunction behind the slit requires at first that the slit is a macroscopic object in the sense that its position relative to the screen is fixed. The wavefunction does not have the same value everywhere on the screen, although it covers the screen in principle completely; the value of the wavefunction at a certain position depends on the position of the slit in the plane parallel to the screen. If we want to measure the displacement of the slider by the passing electron, of course this slider must be freely mobile. Now it is no longer trivial if the mobile slider is at rest, before the beginning of the experiment, or not. We especially have to choose either to define the position of the mobile slider exactly (and hence the slit in it) or its momentum (its state of motion).

When we measure the momentum of the slider after the passage of the electron, the result is only of value for us if we know the momentum it had before, because we want to determine the *change* of momentum by the electron. However, the more accurately we know the initial momentum, the less accurately known is the position of the slider (and thus the slit). If we want to perform the experiment in exactly the same way as before, we have to know the position of the slide diaphragm exactly, thus we cannot determine its initial momentum, neither can we draw any conclusion from a momentum measured after the passage. If we want to draw such conclusions, we have to fix the initial momentum at the expense of the determination of the position. The experiment then cannot run as before, because a wider and flatter distribution of the values of the wave function has to be assumed. The more accurate information on the momentum obtained in this way is devaluated by the fact that we correspondingly know less precisely where the momentum vector begins.

Considering this argumentation, Einstein's next step is astonishing. Apparently we must assume that Bohr mentioned just summarily in the discussion, that the slider has to be regarded as a second quantum object. This again could be understood as if Bohr only regarded the *exact* control of position and momentum of the slide diaphragm as being impossible. So for Einstein the following idea seemed to be straightforward: we only need *minimal* information about the electron, namely if the dislocation is upward or downward. This alone gives more information than the wavefunction delivers. To this end we can modify the experiment such that we maintain the mobile slide of the diaphragm, but now we put a *second* diaphragm between the first one and the screen. This new diaphragm contains *two* slits and the wavefunction passes through both completely symmetrically. By measuring the dislocation of the first slit upward or downward we can detect through which slit the electron passes, although the wavefunction does not contain this information (cf. Bohr (1949), p. 128 f.).

The experiment described here is the well-known two-slit experiment. As far as we know, it originated in the discussion between Einstein and Bohr; the idea most probably came from Einstein. According to the just-described depiction by Bohr, it was used by Einstein to push forward the idea of controlling the electron's momentum — by measuring the momentum of the diaphragm. The double-slit experiment, however, had a side effect on Bohr, which is worth mentioning first. With this experiment it becomes completely clear that the wave–particle problem is not about quantum objects that can appear sometimes as waves, sometimes as particles. We first take a look at the description Bohr gave in a retrospective about the experiment and its course:

> "When a parallel beam of electrons (or photons) [...] hits the first diaphragm, we will observe under normal experimental conditions an interference pattern [...] With intense radiation, this pattern is built from the accumulation of single processes, each of them generating a small spot on the photographic plate. The distribution of these spots follows a simple law that can be deduced from the analysis of the waves. The same distribution should also be found from the statistics over a large number of experiments, that were carried out with such weak radiation that at a single exposure only one electron (or photon) will reach the photographic plate and hit at one point [...]." (Bohr (1949), p. 128.)

Bohr thus clarifies by himself: the double-slit experiment can (at least in principle) be carried out in such a way that there is never more than one single electron passing through the double slit and hitting the screen. After many repetitions of this process, all impact positions on the screen jointly form the interference pattern or wave pattern. The impact position of the single electron, a localized particle, is thus definite (though not completely determined) through a physical process that is described by a wave. This in particular means also that the electron shows in one and the same experiment attributes that come from the two different representations — wave-picture and particle-picture.[12] So the wave–particle problem Bohr originally was

12) One could question the claim that a single electron that hits the screen shows wave attributes, because the wave pattern is formed only through the impacts

confronted with does not exist at all.[13] The fundamental problem of quantum mechanics is a very different one — namely the question whether its descriptions (e. g. of particles by waves) are complete.[14]

What could be a possible objection to Einstein's proposal? One argument is the following. The deflection upward or downward can only be established beyond doubt when the aperture was at rest before. Thus the above objection — the state of motion of the aperture must be known, which goes at the expense of the position of the aperture — can be repeated here.

In summary, Bohr's reply was: of course we can measure the state of motion of the first aperture and thus get information about the trajectory of the electron. But it turns out that with this we have changed the experimental arrangement (because now the aperture is no longer fixed, it must be mobile) in a way that no interference pattern at all can form on the screen. Initially, our attempt was to prove that the behavior of the single electron is determined by both slits. If we were

of many particles and one cannot read in any way from the single impact if it belongs to a wave pattern at all. One can object to this that the position of the impact of a single electron is not totally random, but determined by the wave through probabilities; only in this way can one understand that the position becomes an integral part of the wave pattern. In this sense, the impact position of the single particle is also a wave attribute even when this cannot be recognized from it.

I have mentioned elsewhere (Held (1998), p. 61) the blending of wave and particle characteristics in the double-slit experiment and pointed out that some physics textbooks are in this respect confusing. A clear description of the actual facts is given by Holland (1993), who describes the double-slit experiment first and continues: "When we speak of 'wave–particle duality' we do not, or cannot, mean that matter manifests itself either as a wave in its classical sense or as a particle (again in its classical sense) depending on the experimental arrangement, and never the two simultaneously. The 'wave' is only ever made apparent by observation of particle positions and in this regard has little in common with our classical notion of wave motion. In classical optical or ripple-tank experiments one observes a distribution of interference fringes that is continuous in time. Although we may simulate this structure in quantum mechanics by sending a large number of (noninteracting) electrons through the interferometer in a short time, the pattern demonstrating the wave character is basically granular." (p. 174).

13) Compare to Chapter 5

14) It should be emphasized that Bohr's own retrospective description, which completely clarifies the unification of wave and particle characteristics in an experiment, is from the year 1949. In 1927, when the example was discussed between him and Einstein, this was obviously not yet clear to him (cf., e. g., a statement from 1929 in Bohr (1931), p. 70). From 1935 on, however, the wave–particle problem and the wave–particle complementarity completely disappeared from his texts.

able to prove at the same time that the electron went through one definite slit only, a paradox would arise. We would have to assume (Bohr (1949), p. 129) "that the behavior of an electron [... depended ...] on the presence of a slit in a screen through which it did not go as it can be proved." It turns out now that we can in fact prove through which slit a single electron went, but only by modifying the experiment. As soon as we do this, however, no two-slit pattern occurs on the screen any longer. This means: now, since we know through which slit the electron went, there is no reason left to make any connection with the corresponding second slit. This might be unsatisfying, but at least it does not give rise to any contradiction.[15]

One aspect of both slit experiments, which Bohr and Einstein obviously were not clearly aware of, turns out to be very important from today's perspective. Whether the property Einstein would have liked to get information about really exists was not directly discussed by both physicists and accordingly not questioned by Bohr. Their discussion was just about the question whether such attributes can be determined and Bohr's summarized answer was: yes, but only when the experiment is decisively modified. One could have asked Bohr, if a property that cannot be determined because of the arrangement of the experiment exists at all, and consequently if the single electron, in the case that it participates in the generation of the wave pattern (and therefore its passage through a certain slit cannot be determined), goes through a definite slit. Bohr's words indicate that he would not have considered this question as scientifically meaningful at all, because science is only concerned with whatever is "detectable". The position of the electron in the plane of the diaphragm, however, is not "detectable" when we decide to let it participate in the wave pattern.

It also becomes clear which aspect of such an answer must have remained unsatisfying for Einstein. The apparently plausible limitation of science to whatever can be in principle observed is driven here to an implausible extreme. Whenever we decide to determine the position of the electron in the plane of the diaphragm we can be sure to get an answer. Therefore, the question about the position of the electron in the plane of aperture is, in principle, not meaningless, but it is supposed to be considered as meaningless when we decide not to

15) Compare to Sections 1.5 and 5.1.

answer it. The scientific justification of this question would accordingly depend on whether the experimentalist decides to answer it or not. This means in the end just to avoid an answer all together.

In 1930, three years after these discussions, there was another Solvay conference in Brussels. Einstein, still unsatisfied, had a new gedanken experiment at hand. The special point about this experiment was that Einstein was now attacking a second uncertainty relation, the one for energy and time: the more precisely we know the energy of a quantum object, so the relation says, the less precisely we know the time at which it has this energy (and vice versa). Einstein was arguing here with the theory of relativity and Bohr, after a night of pondering, was also applying the theory of relativity in his turn, in order to disprove that an indirect measurement using an auxiliary object could get around the uncertainty relation.

In his new gedanken experiment, Einstein was looking at a box with a hole that can be opened for a certain period of time, controlled by a clockwork mechanism inside the box. With a beam source that emits photons inside the box, one could achieve that for each opening only a single photon leaves the box exactly at a certain moment, which is determined by the clock. The energy of the photon, however, could easily be determined by measuring the weight of the box before and after the escape and by using the equation $E = mc^2$ from general relativity theory. In this way it should be possible (again using an auxiliary object, namely the box) to determine the exact energy of a quantum object at an exact time — in contradiction with the uncertainty relation for energy and time.

It is not surprising for us that Bohr also knew how to invalidate this example — by consulting the theory of relativity as well. When we assume that the weight of the box is measured with a spring balance, an uncertainty of this weight comes from the fact that we are not able to determine the position of the pointer relative to the scale with any required precision. It is due to the uncertainty relation for position and momentum that the more precisely we try to determine the pointer's position and thus the weight of the box, the less precisely we know the pointer's momentum. The uncertainty of the momentum, however, is directly linked with the period of time that is needed for the whole weighing process: the smaller the uncertainty of position and, thus, the one of mass, the longer is the time interval for weighing. According to the theory of general relativity, a time difference is generated by shifting a clock in a gravitational field. This means, the more precisely the weight of the box is determined, the less precisely the time at which the photon came out is known — in agreement with the uncertainty relation for energy and time.[16]

3.3 The EPR argument as an indirect argument for the incompleteness of quantum mechanics

Einstein accepted being once again defeated. Only afterwards did he realize a peculiarity that distinguishes the box experiment from the previous slit experiments.[17] After the photon has escaped from the box, we are free to choose if we are going to measure its weight and thus determine the energy of the photon or open it in order to read the clock and thus determine the time at which the photon has left. Consequently, we are either able to predict the precise time at which the photon will arrive at a definite far distant position or the energy it has at its arrival. This means, though we are not able to determine both attributes of the far-distant photon at any accuracy, we can still freely choose which one of the two we are going to determine — and in particular with any accuracy — without influencing the photon itself. This version of the experiment cannot help us to outwit the uncertainty relation for energy and time, but it allows us to gain optionally one out of two pieces of information about a quantum object, where the simultaneous determination *would break* the relation. We cannot avoid the uncertainty relation, but the impression is reinforced that properties, which we are never able to determine concretely at an object at the same time, are nevertheless simultaneously present.

This idea — the observer can choose which one of two complementary properties he determines — replaced in Einstein's thinking after

16) See for the calculation Bohr (1949), p. 137 f. and more detailed, e. g., Murdoch (1987), p. 159 and Held (1998), p. 87. The uncertainty relation for energy and time is not as fundamental as the relation for position and momentum (it does not follow from the quantum-mechanical formalism like the latter, because it contains no formal representative for the 'observable time'). Therefore, it is no surprise that Bohr in his argument against Einstein brings the more fundamental relation between position and momentum into the play. One could also consider this as an argument for the fact that the uncertainty relation for energy and time follows from the relation for position and momentum.
17) Cf. Held (1998), p. 98 f., about the historical course.

1930 the direct attempts to prove quantum mechanics as being incomplete. After his emigration to the USA, Einstein tried to exploit this new idea. The result of several years of effort was the above-mentioned article, written jointly with two young physicists and published finally in 1935. The two coworkers were Boris Podolsky and Nathan Rosen; the topics of the essay with the title "Can quantum-mechanical description of physical reality be considered complete?" are — named after the first letters of the names of the three authors — the EPR-argument and the EPR-experiment. The authors try to show that the answer to their question must be No. For this, Einstein's consideration discussed above (the observer can choose which one out of two complementary quantities he observes) plays the central role. In the box experiment we are able to decide afterwards about the determination of one out of two complementary properties, which the now far-distant photon had had before. Through this, the corresponding other property can be predicted with certainty. The authors of the EPR-essay follow exactly this argumentation.

In order to understand the EPR-argument, we now recall the possibility to choose between two measurements, which cannot be carried out simultaneously. If, for a single quantum-mechanical object, two quantities of a suitable kind were simultaneously determined, the uncertainty relation would be broken; if quantum mechanics is a complete description of reality, these properties do not even exist simultaneously. When making use of the free choice of the measurement of one of the two properties, one could always reply the following: the actual measurement is such an intrusion into the quantum system that the measured properties are possibly just *generated* in its course, and the complementary property is deleted at the same time. In this case the argument of the free choice has no effect (depending on the way in which the observer decides and how he actually measures, a property comes or ceases to exist). Therefore, one should arrange a situation where the choice between the measurements of two complementary properties can be made without causally influencing the object. This was already the case in the box experiment. Even there, one cannot argue any longer that the measurement would destroy properties or bring them into being, because the measurement is carried out indirectly at an auxiliary object, which gives information about some other far-distant object without being able to interact with it — unless one assumes that both objects could still influence each other in spite of their distance (namely *nonlocally*). All this, already laid out in the box experiment, is made explicit in the EPR-experiment.

We can translate the experiment into a modern, simpler variation, the so-called EPR–Bohm experiment.[18] A source emits pairs of particles in a special entangled state, the so-called singlet state. The particles fly apart and then they are measured at far-distant apparatuses **A** and **B**. Each apparatus can measure two quantities of the arriving particle, A_1 and A_2, or B_1 and B_2, respectively. The entanglement of the particles in the singlet state expresses itself in the following way: without any further conditions, the measurement value of none of the quantities A_1, A_2, B_1 and B_2 can be predicted with certainty. But when quantity A_1 is measured at apparatus **A** at one particle and the value 1 is found, it can be definitely predicted that at apparatus **B** at the other particle for the quantity B_1 the value -1 will be measured; when for A_1 the value -1 is measured instead, the value 1 for B_1 can be predicted with certainty. (For B_2, however, no definite value can be predicted with certainty.) The same connection exists between A_2 and B_2.

Thus it is possible to measure the value for A_1 or A_2 at apparatus **A** and then to predict with certainty the value for B_1 or the value for B_2 of the far-distant second object. If we assume that only local interactions exist between the particles and the apparatuses, it is impossible that one of the measurements at apparatus **A** causes a change in the proximity of apparatus **B**. Therefore, we are able to predict the value for B_1 or for B_2 with certainty (in principle, to measure it indirectly) without influencing the object. Now comes the crucial additional premise from Einstein, Podolsky and Rosen. The three authors start out from the following principle: *if I am able to predict an attribute of an object with certainty (measure it indirectly) without any influence on the object, then the object has this property.* This principle is used in the following way. With the measurement at apparatus **A** I can decide for example for A_1 and then predict the value of B_1 with certainty, without influencing the particle at **B** and therefore this value of B_1 exists in this case. I can also measure A_2 instead and predict the value for B_2 for sure without influencing the particle at **B**, and therefore a certain value for B_2 exists in this case.

---

18) Compare to Section 2.4.

No matter how I decide — this is the way the authors argue — both cases belong to the "same reality", which means to the same situation, because a decision about a measurement at apparatus **A** cannot directly influence the state of the particle at **B** for the large distance. Quantum mechanics forbids states in which the quantities B_1 and B_2 for one particle simultaneously have values. But since it was shown that there are situations in which a simultaneous existence of such values has to be assumed, the theory is incomplete (cf. Einstein et al. (1935), p. 83 and 86).

Einstein and his colleagues leave it open whether they consider the situation *before* or *after* the choice of a measurement at apparatus **A**. For their argument this is basically irrelevant. Assuming that the measurement of A_1 has already been carried out, the actually predictable value of B_1 is then a true property of the second particle. A_2 could have been chosen instead, so the value of B_2 would have been predictable. Since this choice has no influence on the particle at **B**, the value of B_2 is also real in the case that A_1 was measured. From having, in principle, the possibility of a clear prediction, the authors conclude about the reality of what is predictable — even in the case that this possibility is excluded, because of a real choice against the pertaining measurement. Now we look at the case that none of the possibilities at **A** has been chosen yet. Here, in principle, the possibility of a clear prediction exists for B_1 and B_2. Therefore, also in this case one can conclude about the reality of what was predicted — two definite values for B_1 and B_2.

Bohr reacted immediately and wrote — in the same year 1935, in the same journal, and under the same title — a reply to the EPR-essay. He tried to show anew that the argumentation of Einstein (and his coauthors) was erroneous, that quantum mechanics should be regarded as a complete description of physical reality like before. His argument was focused exactly on the basic principle of the EPR-argument: *when I am able to predict an attribute with certainty (measure it indirectly) without influencing the object, the object has this property.* Bohr considers this principle, when understood correctly, as being acceptable, but in the way it was used from Einstein and his coworkers as "ambiguous" (cf. Bohr (1935), p. 93).

Bohr argues: a physical problem can only be considered to be answerable in a reasonable way, when a measurement situation, which allows finding an answer, is well defined. A quantum-mechanical measurement situation, however, is well defined only after I have made a decision about a measurement setting. Therefore, we are not allowed to consider the situation before the decision about the setting is made, and thus tacitly blend two situations that cannot exist simultaneously. We can understand Bohr as saying that the possibility to predict, with certainty one property through the measurement of another is, in a sense, not sufficient to consider it as being real. This is well intelligible, since the possibility of a sure prediction of an attribute at **B** is linked to an actual choice at **A**; nothing can be predicted without this choice. The actual selection of a quantity is thus for Bohr the condition that a physical measurement situation is well defined, and the situation before the choice of a quantity at **A** is from the outset unacceptable for him. After the selection, a certain one of the two quantities at **B** can be clearly predicted; to this one, according to Bohr, the basic principle mentioned above is applicable. But also the quantum-mechanical description is a different one now — a fact that was pointed out by Einstein, Podolsky and Rosen themselves, without drawing any conclusions. The quantum-mechanical state is no longer the singlet state, from which none of the values of A_1, A_2, B_1 and B_2 was predictable with certainty. It is instead a quantum-mechanical state in which the quantity that can be predicted with certainty at **B** and its value shows up explicitly. The quantum-mechanical description of the now-defined situation is therefore complete.

The EPR essay and Bohr's reply were the climax and the end of the Bohr–Einstein debate. Both physicists now gave up trying to convince the other of their own point of view, and instead refined their own position in further articles. The two essays from 1935 are, nevertheless, the foundation of considerations about quantum mechanics for almost all subsequent authors. Therefore, we should clarify what had been achieved to that point. Einstein, Podolsky and Rosen introduce an *entangled state*[19] for a system of two particles for the purpose of their argument, a state that cannot be considered as being composed from the states of the subsystems. After measuring one of the particles, this state is transformed to a nonentangled state that can be understood as a composed state. This transition is neutrally described by the three authors as "reduction"; if we consider this reduction to

19) Compare to Section 2.2.

be a physical process, we have a further example of the mentioned collapse of the wave function upon measurement, which in this case is the breakdown of the entangled state. The argument of Einstein, Podolsky and Rosen makes it clear that they are not assuming such a collapse. They explicitly start out from the idea that all interactions are local, which means that an influence at **A** can have an immediate effect only on the surroundings of **A**, but not **B**. A collapse triggered by the measurement with apparatus **A**, however, would have an immediate effect on the state of the particle at **B**; the authors assume no such collapse — entirely in agreement with Einstein's original objections twelve years earlier. The peculiarity of the EPR-experiment is expressed in the fact that the collapse is now not just concerning a wave that belongs to a single object, of which the interpretation is unclear — is the wave the object itself or is it just a probability distribution that belongs to it? — but a wave that connects two objects with each other. At the collapse of the wave, *nonlocal* interactions occur between the two particles throwing them from an entangled into a nonentangled state. The fundamental discussion about quantum mechanics thus owes the problem of nonlocality to the EPR-argument.

Bohr's counterposition is to accept the limitation to local interactions and still to consider quantum mechanics as being complete (hence being a complete description of physical situations), without accepting the collapse of the wave function and nonlocal interactions. Unfortunately, it is not clear at all how this should be possible. But to be fair we have to say that Bohr simply refuses to answer the question about the way in which the reduction of the wavefunction should be understood. He does not consider the state before the reduction as well defined, in contrast with the state after it. We have to interpret him favorably here. In particular for entangled states physical measurement quantities and well-defined measurement situations exist, which leave the entanglement intact. Therefore, we have to assume that such a state is not completely undefined, but it is well defined with respect to certain quantities (such that correspond to a measurement at a particle) only with the description of a concrete measurement situation. However, one difficulty that has already been mentioned still remains. We can first produce a certain quantum state of two particles, e. g. an entangled state and then decide what we are going to measure on it (e. g. A_1 or A_2 or also some entangled quantity). When such a decision is the condition for quantities being well defined, and this in turn is the condition that one can speak about a value of this quantity as a true property, whatever about the particles is real depends on arbitrary decisions of the experimentalist. It is not clear in which way Bohr can escape from this difficulty.

Correspondingly, Bohr leaves the answer to the problem of non-locality open. In the case that quantum mechanics is complete, the quantities A_1, A_2, B_1 and B_2 in the entangled state have no definite values. With the measurement of, e.g., A_1, a nonentangled state is formed; in this state A_1 and B_1 have certain values, which in addition are always opposite to each other (as described above). How can it be that through the measurement at **A** not only A_1 but also B_1 gets a value and, moreover, always one that fits with the one of A_1? For the advocate of completeness, there seems to be only the collapse as a possible explanation. Those, however, who do not assume quantum mechanics to be complete must still explain the the perfect opposition of properties. It seems that this explanation inevitably needs to fall back onto hidden quantities.

Bohr does not comment on this problem. He sees the locality condition just as what it is: an auxiliary assumption that creates additional problems but does not concern the core of the matter. He tries to point out that physical quantities are only well defined within a certain context; this is his basic idea. In the EPR-experiment that was forced upon him, these contexts are the actual settings at a far-distant apparatus, but this is just a special difficulty of this experiment. Today, we know that for quantum-mechanical objects the physical quantities are connected with each other in a new, for classical objects, unknown way: the single quantities are *context-dependent*. Either — when we regard quantum mechanics as complete — the quantities depend on each other concerning the question whether they have values at all and if so, which ones; or the quantities are — when we regard quantum mechanics as incomplete — at least dependent on each other concerning the values they have. Bohr's way, insisting that physical quantities are only well defined within certain contexts, i.e. the measurement contexts, can be understood as an intuitive insight into this context-dependence.

Even Einstein realized that the actual question about the completeness of quantum mechanics had basically nothing to do with the locality problem. In his correspondence with Erwin Schrödinger in summer 1935 he again only considered single systems and the question whether their quantum-mechanical description is complete. The two physicists discussed macroscopic examples for illustration and Schrödinger devised his famous cat; connected to a quantum object it became a quantum object by itself. Since the cat was set in a superposition of the states "alive" and "dead", it became impressively clear what it means to assume that a quantum-mechanical quantity sometimes has no value at all, and that it gets its value in the course of a measurement.

3.4 Nonlocality, entanglement and dependence on the context

The efforts of physicists concerning the fundamental questions of quantum mechanics have, in principle, reproduced the last step of the Bohr–Einstein debate. John Bell was able to show with the inequalities[20] named after him that the EPR-experiment can be realized; corresponding experiments have been carried out and have confirmed quantum mechanics but disproved theories of hidden quantities (as far as they are not context-dependent). Bell himself, however, made considerations according to which the breaking of the locality requirement is pushed into the background and the context-dependence of quantities or properties comes to the fore instead.[21] Today it can be shown that for a single quantum object not all quantities can have well-defined values that are context-independent.

The problem of nonlocality turns out indeed as being derivative, as one that follows from the structure of quantum mechanics, when certain systems (of several particles) in certain (entangled) states are considered, and in addition the particles are assumed to be far apart from each other. First it is true that quantum-mechanical quantities and properties are context-dependent in the explained sense. This context-dependence can be demonstrated for a single particle; then, however, the number of necessary quantities is quite large. Simpler and more elegant demonstrations use quantum systems of two or

20) Compare to Section 2.5.
21) See regarding the inequalities Bell (1987), p. 14–21, about context-dependence Bell (1987), p. 1–13, especially 6–9. The latter essay is from 1966, and it contains the basic idea of the so-called Kochen–Specker theorem. For explanations see Redhead (1987), p. 27–30 and 119–138.

three particles, but here it is not necessary to assume that the particles are far apart. In such a system, there can be entangled states, and in such states the entire system has properties, that cannot be simply traced back to the ones of its parts. There are "entangled quantities" that have values, although the quantities they consist of have no values by themselves. When systems in such states are "pulled apart", i. e. the components are separated by a large distance, nonlocal entanglement of the states (nonseparability) occurs, and for the measurements at the single components there are nonlocal couplings of the results.

The debate between Bohr and Einstein leads initially towards the problem of nonlocality (in the form of the EPR-argument) but on a closer look also away from it again. The fundamental problem behind nonlocality and the possibility of entangled states is the context dependence of quantum-mechanical quantities and attributes. This context dependence has its roots in the structure of quantum mechanics, a structure that has not yet been entirely clarified but was postulated because it describes the atomic objects correctly. We can accept this structure, and the problematic properties, which follow from it, as given, but this would neither do justice to Einstein's standards for scientific explanations nor to Bohr's efforts to give such explanations for quantum mechanics. This could only be achieved from an explanation of the structure of quantum mechanics and its context-dependence.

References

- K. Baumann, R. U. Sexl (1984) (eds.): *Die Deutungen der Quanten-theorie,* Vieweg, Braunschweig and Wiesbaden.
- J. S. Bell (1987): *Speakable and Unspeakable in Quantum Mechanics,* Cambridge University Press, Cambridge.
- N. Bohr (1931): *Atomtheorie and Naturbeschreibung,* Springer, Berlin.
- N. Bohr (1935): "Can quantum-mechanical description of physical reality be considered complete?", *Phys. Rev.* **48**, 696–702.
- N. Bohr (1949): „Diskussion mit Einstein über erkenntnistheoretische Probleme in der Atomphysik", in P. A. Schilpp (ed.): *Albert*

Einstein als Philosoph und Naturforscher, Kohlhammer, Stuttgart 1955.
- *Niels Bohr. Collected Works* General Editors: Erik Rüdinger/Finn Aaserud. 10 volumes. North-Holland, Amsterdam 1972 ff.
- A. Einstein, B. Podolsky and N. Rosen (1935): "Can quantum-mechanical description of physical reality be considered complete?", *Phys. Rev.* **47**, 777–780.
- C. Held (1998): *Die Bohr-Einstein-Debatte. Quantenmechanik und physikalische Wirklichkeit,* Schöningh, Paderborn.
- P. Holland (1993): *The Quantum Theory of Motion,* Cambridge University Press, Cambridge.
- R.L. Liboff (1997): *Introductory Quantum Mechanics,* 3rd edn., Addison-Wesley, Reading, Mass.
- D. Murdoch (1987): *Niels Bohr's Philosophy of Physics,* Cambridge University Press, Cambridge.
- M. Redhead (1987): Incompleteness, Nonlocality and Realism. A Prolegomenon to the Philosophy of Quantum Mechanics, Clarendon Press, Oxford.

4

An excursion into the quantum world

Robert Löw and Tilman Pfau

4.1 First steps in the quantum world

Would you like to have a cake and eat it at the same time? This is impossible? The quantum world can manage that! How come? In here, a particle can become a wave that is oscillating in many places simultaneously. This, of course, is in contradiction to our everyday experience, because every attempt at having a cake and eating it must fail. Material waves can, nevertheless, achieve this without any problems. They can oscillate in many places simultaneously like, for example, a water wave, which is also in principle not limited to one certain position: it can split and run simultaneously through two channels.

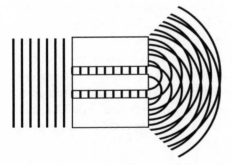

Fig. 4.1 A water wave can simultaneously run through two channels. In the quantum world, a single atom is able to do this as well.

Anyone with a passion for murder mysteries would sooner or later raise the question, if this could make the perfect murder possible.

Entangled World, Jürgen Audretsch
Copyright © 2005 WILEY-VCH Verlag GmbH & Co. KGaA, Weinheim
ISBN: 3-527-40470-8

When a potential murderer is present at the site of crime and somewhere else at the same time, he could actually have an unassailable alibi! Fortunately, not even quantum mechanics allows for the perfect murder. Even though a criminal quantum-mechanical particle could really be in two places simultaneously. The particle has to interact with the victim at the scene of the crime in order to commit a murder. By doing so, it reveals its whereabouts. A quantum Sherlock Holmes can therefore leave a sigh of relief, because the quantum-mechanical wave is necessarily changing through the interaction — the deed — back to a classical and thus localized particle, and the alibi vanishes into thin air.

How this process of transformation from spatially distributed waves to particles tied to a certain place and back proceeds, will occupy us throughout this chapter. Where does the transition between quantum mechanics and the classical world happen? Is it the act of murder by which the culprit reveals himself, is it the victim that forces the culprit to be in a definite place, or is it the surrounding world that establishes the location of the crime by observing the murder? Here, the concept of entanglement comes into the discussion. We will see that entanglement is an everyday and everywhere-present phenomenon. Entanglement is the reason why our day-to-day experience is shaped by classical phenomena despite the quantum-mechanical core of the matter. Our world is therefore naturally an entangled world.

Entanglement exists around us in such wealth that any control over it appears to be a forlorn hope at first. The challenge for science nowadays is seen exactly in the controlled creation and manipulation of entanglement in manageable systems. With this, we might succeed to find our way into the complexity of the quantum world and even to make good use of its peculiarities. Examples in this direction are the quantum computer, quantum cryptography and quantum teleportation, of which the classical counterparts, provided that they exist at all, are less efficient in many respects.

4.2 On the history of the quantum theory

One hundred years ago, a world in which matter behaves like waves, or an object that appears to be in different places at the same time, was not conceivable at all. The physics consisting of mechanics and electrodynamics was in best harmony with nature in the way it was visible and could be experienced every day. All problems of physics appeared to be solved. According to the general opinion, still unexplainable phenomena were expected to be deciphered sooner or later with a more consequent application of the classical laws of nature. The canon of physics appeared to be concluded.

Those last and apparently small problems besieged by scientists were nevertheless not willing to give up their antagonism. It was due to the understanding of the color spectrum of a body glowing from heat that a completely new direction came into view. In the year 1900, the solution of this problem led Max Planck to the quantum theory, which kept us in suspense to this day. Planck was formulating the condition that light cannot be emitted in arbitrarily small energy packages, but only in the form of energy packages of a finite size — the photons. Planck was originally absolutely unsatisfied with his solution, so that he introduced a quantity h, which he at first named an auxiliary quantity (German: Hilfsgröße h). It is exactly this quantity h that symbolizes a single energy package, and that is today known as Planck's constant (German: Wirkungsquantum). This was the genesis of quantum theory.

At the same time models of atoms were developed, which were also not within the framework of classical physics. The basic structure of the hydrogen atom (consisting of nucleus and electron) for example had been known for some time. Also, the mass and the charge of the single components could be determined with a fair precision. Since the mathematical form of the attractive force between electron and nucleus was exactly equivalent to the force between sun and earth, the motion of the electron on elliptical orbits around the nucleus could be imagined without great difficulties. Electrodynamics and Newtonian mechanics were accepted as watertight theories; but they were not able to explain the fact that the hydrogen atom is stable. An electron on an elliptical orbit is nothing but an antenna that is permanently emitting energy. As a consequence, the electron must collapse into the nucleus within the shortest time. The loophole from that dilemma — it was already well known that hydrogen atoms are quite stable — came from the revolutionary approach of quantum mechanics.

Quantum mechanics was developed in the first half of the 20th century and it initiated one of the great revolutions of the scientific view of the world. For its discoverers it was the key to many unsolved physical phenomena, but opponents dismissed it as complete nonsense. It appeared to be obscure, not understandable or even wrong not only to outsiders but also to acknowledged physicists. The advocates of quantum mechanics were trying to comprehend the theory with gedanken experiments, today they would be called virtual experiments, which became famous. Real experiments failed during that time very often because of the available experimental techniques. Particles that can be in two places at the same time, matter that behaves like waves, particles far apart from each other while the one still "knows" about the state of the other. In the meantime, such mysterious theoretical constructs were realized in experimental setups, and it turned out that the astonishing predictions with all their contradictions were actually true. Are these really contradictions? How can something simultaneously be in two places? We think of the laws of quantum mechanics as being contradictive, because they appear to be incompatible with the world of our everyday experience.

We experience the effects of natural laws day-by-day with our senses and we construct from that our own edifice of empirically established figures. This is why Newton's law of gravity immediately appears to be plausible to us. When an apple falls from a tree, the gravitational force can be seen by eye. The fact that the earth is rotating around the sun is already a bit more complicated, but still comprehensible after some consideration. Perhaps it is helpful to visualize this situation with the example of a merry-go-round: the centrifugal and gravitational forces exactly balance each other. These effects, which we observe in nature, are called classical in the sense that they can be understood with classical mechanics and electrodynamics. To be able to understand quantum mechanics, we have to leave the classical world.

An electron "rotating" around the nucleus can simultaneously be in several places in the quantum world. This is different in the classical world (sun and earth): the "particles" involved in this process (celestial bodies) are at a certain time fixed to a certain place. An electron, which is bound to a nucleus in the quantum world, is described by a matter wave with its oscillation spread around the whole nucleus, and is therefore present in many places. This property, as an abstract image, is not so easily accessible. However, was it not inconceivable to consider the earth rotating around the sun not too long ago?

How does the quantum world become visible to us? It is possible now to record the wave character of matter as "photographs" in experiments. Fig. 4.2 shows an electron that is sloshing around as a wave in some kind of a microscopic bathtub.

Fig. 4.2 To the left we can see a classical particle, for example a billiard ball on a round billiard table. When the ball is given a push and friction is neglected, the ball will once roll over each place on the table after lots of collisions with the cushion. At any time the ball will certainly be exactly in one definite place on the table. To the right we see the same system, but with a billiard table that was fabricated a hundred million times smaller. The "cushion" consists of 74 iron atoms, which were strung together on a very flat copper surface. Using an atomic force microscope, it is possible to recognize single atoms, pick them up and deposit them again at a certain spot. The billiard balls in this case are the electrons. The pictures visualize the probability of the electron to be in a certain place. The higher the elevation, the larger is the electron density. The wave-type structure inside the iron ring shows that the electrons do not take exactly one possible position, but they are spread out over the whole "table" — however, not everywhere with the same probability (source: www.almaden.ibm.com/vis/stm/).

The fact that even atoms and molecules can behave like a wave can be shown with an atomic interferometer. Waves have the characteristic ability to interfere with each other. With an atomic interferometer, the interference of matter waves can be visualized.[1] For this, it is first necessary to understand the meaning of interference.

Interference is an effect that occurs with the superposition of at least two waves. The waves can amplify each other (constructive interference) or cancel out each other (destructive interference). The

[1] Compare to Section 1.3.

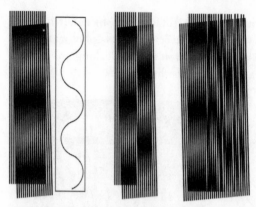

Fig. 4.3 Constructive and destructive interference. The superpositions of two plane waves (waves with a linear wave front) are shown. Depending on the positions they have relative to each other, they either amplify or cancel out each other. The interference can be recognized on the left side as vertical modulations (almost horizontal fringes). The pattern is very sensitive towards small changes of the relative phase between both waves. In the middle, a single wave front was shifted by just half a wavelength. Due to this, the interference pattern is also shifted by half a wavelength. When the shifts between the wave tops show a stronger variation, the interference fringes disappear completely.

visibility or the contrast of the interference fringes formed is a measure of the degree of the wave character of the original single waves.

When trying to observe interference fringes for physical particles, first two waves that are able to interfere have to be generated. In an atom interferometer (see Fig. 4.4) atoms are shot at a double slit, which splits the atom into two matter waves. The interferometer itself consists, in principle, of a diaphragm with two small apertures and a screen for the detection of the atoms. From the left, a single atom reaches the double slit as a plane wave. Then the one atom passes through both slits simultaneously, in the same way as it would, for example, with a water wave.

Behind the diaphragm a spherical wave spreads out, one coming from each of the two slits. Two waves that can interfere are formed like this. At first it appears to be somewhat unusual that an atom can interfere with itself, but quantum mechanics makes it possible. On the screen, however, only one single "point-shaped" atom is registered. So the particle is not spread out homogeneously to form an interference pattern, but many atoms must have passed the double slit in order to form a recognizable pattern.

Of course now one would like to know if the atom really flew through both slits simultaneously. For this purpose we first close one of the two slits to make sure that the atoms can only pass through the other slit. The interference pattern disappears! This can easily be understood, since at one slit only one spherical wave is generated but not two. Now let the atom fly through the double slit and try to find out immediately behind the diaphragm through which one of the slits it passed. For this purpose, we align a laser beam at a short distance behind the diaphragm, which makes the observation of the position of the atoms possible by fluorescence. Once again, the interference pattern disappears! Initially there are two spherical waves, but as soon as we take a look where the atom is, it is forced to admit being in one place. In this way it becomes a localized particle again, which cannot show interference any longer. This procedure is called the measurement process in quantum mechanics. A quantum system can be in several states simultaneously, until one tries in fact to realize its state. At this moment the wave function is changed, so that the system occupies only one of the available states. This is also why the perfect quantum murder is impossible: at the moment when the culprit acts, he interacts with the victim and can be localized at the scene of the crime.

When the same experiment is carried out on a macroscopic scale, for example by shooting billiard balls through a grating of corresponding size, just a regular distribution of the balls is obtained. Why does the microscopic particle behave like a wave but not the macroscopic billiard ball? When taking a look at the dimensions of Planck's constant h, one realizes that h is a product of a distance, a mass, and a velocity. Now the numerical value of h is extremely small. It can be determined experimentally very accurately, but compared to the length, mass, and velocity of objects around us, it is almost insignificant. Meaningful investigations are carried out with small, very light and very slow objects in order to visualize the quantum world. Particularly suitable are elementary particles like electrons, which are many times lighter than atoms.

Instead of observing electrons with the best microscopes in the world (see Fig. 4.2), one can also try to visualize quantum mechanics

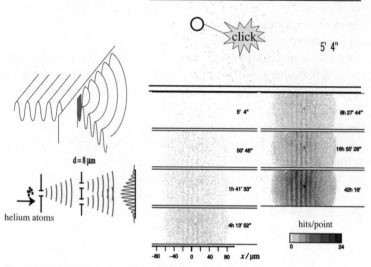

Fig. 4.4 Atom interferometer. A beam of helium atoms is sent through a double slit (left). Each single slit is a source for a matter wave. The two matter waves formed can interfere with each other constructively or destructively. The beam is now attenuated so much that there is always only one atom on the way. Interference arises at the superposition of the two matter waves; just one single particle is detected each time as a dot on the screen. The typical fringe pattern can be recognized (right) only with the statistics after many atoms passed through.

on larger scales. The wave character of atoms can nowadays be observed with a normal light microscope provided that the atoms are cold enough. We have seen that Planck's constant is also proportional to the velocity, which for its part is connected with the temperature of the particles.

For an access to the quantum world, the atoms have to be slowed down a lot. When the mean velocity of the particles in a gas is lowered, the temperature of the gas falls. When we succeed to slow down the particles by so much that the product of mass, velocity, and spatial extension becomes comparable to Planck's constant, the quantum attributes will become visible. The information about the wavelength of a particle is given by the de-Broglie relation

$$\lambda \propto \sqrt{h^2/mT} \tag{4.1}$$

The smaller the mass m or the temperature T of a particle, the larger is its wavelength λ (h is again Planck's constant).

The most recent advances in the generation of ultracold atoms are here of particular relevance. Only the development of completely new types of "refrigerators" could drive the research on macroscopic quantum systems forward at an unusual speed.

Fig. 4.5 The Nobel Prize in physics 1997 was awarded to S. Chu, W. D. Phillips and C. Cohen-Tannoudji for their pioneering studies on laser cooling.

4.3 Cooling techniques and atom lasers

The reduction of the particle velocity, as explained above, provides an important access into the quantum world. The kinetic energy of the particles in a gas (and together with this its temperature) has a quadratic dependence on their mean velocity. A gas that is colder by a factor of four therefore consists of particles with half the velocity on average. When a gas of room temperature (about 300 K) is cooled down by a factor of 10^8 to 3 μK, the particles will slow down by a factor of 10^4. At the same time, their matter-wavelength will grow by the same factor (see Eq. 4.1). The wave properties can then already be observed with a simple light microscope.

Such cold gases, however, cannot be produced with conventional techniques. In our everyday life, we can clearly perceive a cooling of a few per cent, which for example is achieved with a refrigerator. For larger temperature steps, special refrigerators are necessary. About a factor of 4 (relative to the room temperature) can be cooled with liquid nitrogen, about a factor of 100 with liquid helium. When the helium technique is perfected with the so-called dilution cryostat, one can get down at least to the millikelvin range. This is already connected with an enormous expenditure on equipment. Therefore the development of laser cooling caused a sensation. With that, one could immediately venture into the microkelvin range.

The principle of this method is at first astonishing. When a gas captured in a vacuum chamber is irradiated with a laser, it cools down to only a few microkelvin within a few thousandths of a second. When we place a finger in the laser beam instead, we are in the best case burning ourselves — and it is well known that laser beams are also used for welding! Then how can a laser beam cool? We have to investigate the interaction between atoms and light in more detail to be able to understand this.

When irradiated with light, atoms can be transferred from the ground state into excited states, by absorption of energy from the beam in terms of a light quantum. In the course of the collision with this photon, a transfer of momentum to the corresponding atom occurs as well, by which the velocity is changed.

The absorption of light is always followed by an emission. However, the light happens to be emitted in a random direction, which will not become noticeable on average as a change of velocity of the particles. After the emission, the atom will be back in the ground state and it can again absorb a photon from the irradiation direction. The overall force on the atoms corresponds to a change of momentum per time unit and is thus directly proportional to the absorption and emission rate. The emission rate is always the same for a specific kind of atoms. The absorption rate has to be manipulated in order to preferentially decelerate the fast atoms with the light force while leaving the slow ones unaffected. The larger the rate, the better the light frequency agrees with the characteristic frequency of the atom.

For laser cooling, the laser has a particular frequency that does not exactly match the resonance frequency of the atoms but is slightly lower. Immobile atoms are not efficiently excited with this frequency. When the atom moves towards the light source, this frequency is slightly shifted upwards from the "point of view" of the atom and thus towards its resonant frequency. This so-called Doppler effect is known from everyday life. The pitch of the sirens from an ambulance appears higher to us when the car is approaching, and lower when the car is departing. In this way the atoms moving towards the light source absorb the light more efficiently than atoms moving in any other direction. The atoms coming towards the light are at the same time particularly efficiently decelerated, because the momentum direction of the arriving photons is opposite to the direction of motion. This is the secret of laser cooling. When pairs of antiparallel laser beams are sent from all three spatial directions into an atomic cloud, the atoms cannot move out of their resting position — otherwise they would immediately be caught by a laser beam and brought to a standstill.

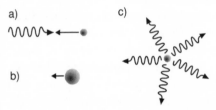

a)

b)

c)

Fig. 4.6 Laser cooling. (a) A hot and therefore fast atom flies from right to left. The velocity is indicated by the length of the arrow. A laser beam (wavy line) comes from the left. The photons in the light field of the laser have a constant momentum, which is transferred to the atom with the absorption (b) The atom has a lower velocity now, but it is in an electronically excited state. The atom releases the excitation energy again to the surroundings in the form of a photon. Now it can absorb the next photon. The emission, however, occurs spontaneously and in a random direction (c), so that the momentum transferred to the atom is averaged to zero after many absorption and emission cycles. Only the momentum transfer from the absorption still contributes, because it always comes from the same direction. In this way the atom is effectively decelerated and therefore cooled down.

Of course this so-called Doppler-cooling technique also has its limits (approximately at 100 microkelvin), and the reasons mainly come from the random emission processes. With more tricky arrangements, which use the polarization of light, temperatures of a few microkelvin can be achieved. The atoms then move only by a few centimeters per second.

A gas can be cooled this much only with a perfect isolation from the environment. Any contact with a (comparably) "hot" wall of a thermocontainer would immediately heat up the gas. Even liquid helium is a thousand times hotter than such atoms. A magnetic field for example can provide an isolation layer. Floating between two strong magnets, gases can be kept in a vacuum chamber for several minutes without getting significantly warmer. This amount of time is then available for experiments.

However, should gases at such low temperatures not actually become frozen to a solid? In fact this will happen, if the gas is given enough time. Normally the density of the gas is so low that on the way to the solid state molecules are formed only very slowly and immediately leave the trap. The laser cooling causes a quick freezing and thus a strongly undercooled gas is obtained.

For temperatures of a few microkelvin, the matter wavelength already amounts to a fraction of a micrometer. With this, however, the experimental art is not at its limits yet. When the cold atom cloud is kept in the magnetic trap, the atoms can be cooled further down by letting the "hottest" atoms "evaporate". This principle is equivalent to the cooling of a cup of coffee. We blow the vapor away, and by doing so we remove water molecules that are hotter than average, which causes the reduction of the average temperature of the remaining molecules. With this evaporation cooling, gases can be cooled down to a few nanokelvin and below. The matter wavelength then arrives at the two-digit micrometer range. Now, it is no longer so difficult to observe the wave properties directly with an optical microscope.

Two fundamentally different kinds of particles are distinguished in the quantum world, the bosons and the fermions. The details of this classification, which is based on the symmetry properties of the wave function, are of no relevance here. It is sufficient to know the following: at low temperatures bosons and fermions behave in completely different ways. In contrast to fermions, bosons can all together occupy the lowest-lying available energetic level. Therefore, when a gas that consists only of bosons is cooled in the way described above, all particles unite to one single large symmetric wave state at a certain temperature. In the case of several million atoms, one speaks of a Bose–Einstein condensate. The Nobel Prize in physics was awarded to E. Cornell, W. Ketterle, and C. Wiemann (Fig. 4.7) for the first experimental realization of such a Bose–Einstein condensate.

We have seen before that single atoms can interfere with themselves. Also, a single giant matter wave formed of many bosons in the same state should be able to interfere with itself. Researchers have

Nobel prize 2001

Cornell   Ketterle   Wieman

Fig. 4.7 The Nobel Prize in physics 2001 was awarded to E. Cornell, W. Ketterle and C. Wiemann for the first experimental realization of a Bose–Einstein condensate in atomic gases.

"cut" a single Bose–Einstein condensate with a laser into two parts. These condensate parts can be brought to interference in the same way as the spherical waves that are formed when atoms pass through a double slit. When the confining magnetic trap is suddenly switched off so that the condensates can freely expand, both parts start to overlap and regions of constructive and destructive interference become visible. A fringe structure like the one we have encountered in the atom interferometer can clearly be seen in Fig. 4.8.

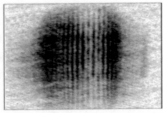

Fig. 4.8 Interference of two Bose–Einstein condensates. A single Bose–Einstein condensate was cut with a laser into two parts. The two condensates (left) are about 0.1 millimeter long. When the condensates are released from the trap, they start to expand and to superimpose. In the superposition area interference occurs between the two matter waves. The resulting interference pattern is shown on the right side. (Photos: *Science*, January 31, 1997.)

A system related to the Bose–Einstein condensate is the common optical laser. All photons have the same frequency and a certain phase relation with each other — so they also occupy one single wave state with a high population. The difference, however, is that the Bose–Einstein condensate rests in a trap while the laser radiation propagates in space. In analogy to the optical laser, a matter-wave laser can indeed be realized. For this purpose, a hole is drilled into the trap that encloses the Bose–Einstein condensate so that the atoms can "flow" out. In the condensate and therefore also in the decoupled beam, the atoms all have the same phase.

To date, a multitude of optical elements for matter waves has been developed with an analogous functionality to their counterparts constructed for light. Mirrors, lenses, beam splitters, waveguides, and amplifiers are available now, and all classical optical experiments can in principle be carried out with matter waves. The advantage of matter waves in contrast to light optics lies in the incomparable precision of the former. For example, the presently most sensitive rotation sensor is based on matter-wave interferences. Matter waves can also be used to clearly illustrate the phenomena that are typical of the quantum world in the way that the gedanken experiments of the grandfathers of quantum mechanics are reproduced.

4.4 Coherence and entanglement

The coherence and the entanglement of states are important concepts of quantum mechanics. They form the basis for spectacular applications like quantum cryptography and quantum computers. The lifetimes of entangled states generated in a controlled way in the lab, however, remain quite low, at least compared to the time of arithmetic operations, which are supposed to be carried out in quantum computers, for example.

What is hidden behind the terms coherence and entanglement? How can coherence arise or disappear? In order to understand this, we will once again go back to the interfering waves. When our waves are nicely formed, we can see on the screen — as discussed above for the double-slit example — a stable interference pattern with regions of amplification or annihilation. The fringe pattern is therefore stationary. When one of the waves is retarded against the other, the fringe pattern is shifted, in fact by one fringe period when the relative shift of the wave is exactly one wavelength (see Fig. 4.3). When this relative shift is varied in a random way, the fringe pattern jumps to and fro until an averaged fringe pattern becomes visible. The contrast of this pattern diminishes when the random shift length is increased.

Our interference image at the double slit is formed now by many single experiments. One atom after the other is sent through the double slit. A clear interference pattern is only obtained when all the atoms have an equal phase — otherwise, each atom will try to form its own interference pattern. In this case, the sum over the single experiments no longer shows interference. Distortions in the phase relation can, for example, occur on the way from the double slit to the screen, when the atoms are not sufficiently isolated from the surroundings.

The meaning of coherence is plainly the ability to generate interference. All processes impairing this are called decoherence.[2] Quantum-mechanical waves lose their coherence not only by random influences from outside, but also by each attempt aiming at the determination of their place. In the course of this process, an entanglement between the quantum object and the measuring instrument occurs, from which an observer for example wants to read through which one of the two slits in the double-slit experiment the atom was flying. We are going to discuss this now in more detail.

When an atom is prepared in a wave state running through two slits (1 and 2), it is in principle not known through which slit the atom is really flying (see Fig. 4.9). It flies with the same probability through slit 1 or slit 2. When the position of the atom at the slit is measured, the atom appears with an equal probability at each slit. However, for a matter wave that runs without a measurement over a distance onto a screen, the probability distribution for the impact of the atom shows interference fringes. So the probability for the atom to hit a certain position on the screen can be predicted (it is surely never going to hit on absolutely "dark" areas). This prediction can only be tested when many atoms prepared the same way are sent through the slits and the results on the screen are summed up. Most of the atoms hit on the bright fringes, and only a few on the dark fringes of the interference pattern.

We have already mentioned that a measurement of the localization of the atoms causes a distortion of the wave and thus the disappear-

2) See Chapter 8.

or

Fig. 4.9 When a photon that was scattered from an atom
at the double slit is observed, it comes from the upper slit
when the atom was at the upper slit or from the lower slit
respectively, when the atom was there. An entangled state is
formed due to the coupling of the two systems, photon and
atom.

ance of the interference pattern.[3] We are going to follow this process
in detail in order to clarify the reason for this. The light beam that
irradiates the atoms behind the double slit for the purpose of finding
out which slit the atom has passed cannot be infinitely weak: at least
one light quantum has to be deflected by the atom, and afterwards it
carries the information about the position of the atom. Therefore, at
least one photon is scattered by the atom during the irradiation. In
the course of this process, the previously unknown information about
the position of the atom is transferred to the photon. An entanglement between the states of the atom and the photon is formed. This
means the following: when an atom comes through slit 1, the photon
is emitted from 1 and vice versa. Immediately after the collision process, as long as the photon has not been measured or detected, neither
the position of the atom nor the position of the photon is revealed,
but both positions are linked with each other. When one of the two
particles is observed at slit 1, it is known for sure where to look for
the other one. This correlation is formally described by an entangled
state.

Correlations appear both in the classical world and in the quantum
world. They are actually something very ordinary. The principle of
the classical correlation was described by J. Bell in his famous example of Bertelsmann's socks as follows. Dr. Bertelsmann, as everybody
knows, always puts on a red sock and a green sock in the morning
without looking. At first it is totally unclear on which foot he is wear-
ing the green or the red sock. As soon as an observer sees that the
green sock is on the left foot, he knows for sure, without having seen

3) Compare with Section 5.1.

it, that the red sock is on the right foot. The "measurement" happens at the moment when the first foot comes into sight. The state of the two socks, however, is classically correlated. The red sock contains the (position) information about the green sock and the other way around.

Fig. 4.10 Bertelsmann's socks.

This is very simple. Nobody would suspect here anything mysterious. Under certain conditions, however, an entangled quantum state shows stronger correlations than what is possible in the classical world, because here the correlation exists already for the wave amplitudes. The probability correlations result from the absolute square of the amplitudes. Due to constructive interference of the correlated wave amplitudes, stronger correlations can result than in the classical case where no interferences occur.

The entanglement between the atom and the photon is terminated for example at the moment when the photon is measured (absorbed). With this, the photon transfers the information about the whereabouts of the atom to the surroundings, the atom is localized and the interference pattern or, respectively, the coherence disappears.

There are several possibilities to explain the disappearance of the interference pattern. For example, the mechanical model: in the course of the emission of the photon, a random momentum is transferred to the atom, which leads to a random change of motion and thus a deferment of the atomic wave. The interference pattern becomes blurred and when averaged over many atoms it disappears. The contrast of the interference pattern is therefore a measure of the decoherence.

The situation discussed here also holds in general. A complex system that is not isolated from the surroundings is permanently entangled by interaction with these surroundings. Light scattering is only one of the possible mechanisms that are relevant here. In our classical world we make no use of the information, but, nevertheless, entanglement occurs everywhere around us. The decoherence, which is linked to this, leads to the fact that we experience our surroundings as a classical world.[4] When we want to experience the quantum world and control the entanglement, we have to prepare systems that are perfectly isolated from the surroundings. There, we can slowly learn how to master the complexity of the entanglement and use it for new applications.

4.5 New dimensions in the quantum world: the quantum computer

Computers surround us in our professional and private lives. The progress of the processor technology keeps us all in suspense. Even though the function of the processors mainly relies on the quantum physical properties of solid-state silicon, they work classically. They process the information incurred in the form of classical bits (0 or 1). An increase of performance under these conditions can only be achieved with higher clock speeds, parallel processing, with several processors or an advanced miniaturization. Natural limits are set for the performance of such kinds of computers. A miniaturisation of the components is worthwhile, because the information cannot spread out faster than the speed of light. The larger the processors, the more time is needed by the signal to get from point A to point B inside the chip. For this reason, an intense effort is nowadays dedicated to innovative components, for example a single-electron transistor. Information that is represented by only one electron is the utmost limit of what can be achieved. These transistors react very sensitively to their surroundings. When the electron is "lost", the whole infor-

---

4) Compare to Chapter 8.

mation is irretrievably gone. Therefore, such components have to be very well isolated from their surroundings, for example neighboring elements, which will probably limit their size to the length scale of several hundred atoms. Smaller building elements cannot be produced for reasons of principle.

There are also other possibilities to improve the performance of computers. The present computers work with very simple logic compared, for example, to the human brain. The reason for this is the enormous networking of the neurons in the brain. Inside the computer two bits are always combined to a new bit by a logic operation. The single arithmetic steps are processed in a sequence, as on an assembly line. In contrast to this, the information in the brain is branched many times and thus it can interact in parallel with a multitude of other information. A brain operates highly nonlinearly. Meanwhile, some research on nonlinear architectures for classical computers is going on, but the distinct complexity of these systems is also causing severe difficulties.

A completely new approach in the field of data processing is the quantum computer.[5] This type of computer no longer calculates with the classical bits 0 and 1, but with quantum-mechanical states instead. The following example will illustrate this.[6]

In analogy to the bits 0 and 1, we name the slits of an atom interferometer state 0 or state 1, respectively. The matter wave, from the quantum-mechanical point of view, can fly through both holes simultaneously. Therefore, the quantum-mechanical state of the atom behind the slit is sometimes neither 0 nor 1, but a superposition of both. These states are also called qubits. In principle, any quantum system that consists of two states, which have to be selectively addressable, could be used for the physical realization. Systems to be considered are, for example, the electronic excitation states of ultra-cold ions or atoms, polarization states of photons, or the alignment of magnetic dipole moments of electrons or atomic nuclei. Being able to generate and read the single states, though, is not sufficient; the possibility to form an entanglement between both qubits in a controlled way and to carry out an arithmetic operation must also exist. Controlled entanglement leads to fast-growing quantum correlations, which can be used efficiently to solve complex problems.

5) Compare to Sections 6.2, 7.4 and 9.1.
6) See also Chapters 6 and 7.

The realization of a quantum computer is still only a dream, but there are different experimental and theoretical groups worldwide working on concepts and their implementations and first simple calculations are already accomplished. What should the particular use of quantum computers be, provided that they come into being one day? For this, we are going to consider a few examples.

Think about a telephone book with 2 million entries. The problem to find the correct participant to a given number would require about 1 million search steps for a classical computer. A quantum computer could do this job in about 1414 search steps, provided that the incurring data are stored in a quantum-mechanical register.

Another possible application lies in the field of secret data transmission. Modern encoding procedures rely on the fact that each classical computer can easily form the product n of two very large prime numbers p and q. The opposite procedure, namely to find the prime factors of a number n is, however, a very difficult problem, which in addition consumes an *exponentially growing* amount of time, the larger n is. The classical computer checks for each single number if, by any chance, it is a factor of n. This takes an extremely long time, and the coding keys are correspondingly safe. Quantum computers, on the other hand, can be far superior to classical computers by using the entanglement for such complex applications.

So are we going to have a quantum computer for the household one day? The following comment has to be added here: most of the algorithms our present calculators work with will not become simpler by using a quantum algorithm; hardly any time can be saved. A quantum computer offers advantages in principle for special problems. The development is still in its infancy and therefore one should be prepared for surprises.

4.6 Approaches for the quantum hardware

Meanwhile, many groups of experts all over the world are working experimentally and theoretically for the realization of quantum computers.[7] The first experiments of the demonstration of quantum algorithms have been performed. For this, it is crucial that the quan-

---

7) Compare to Chapters 6 and 7.

tum particles (qubits) are well isolated from the environment, in order to avoid any uncontrolled interaction with the surroundings.

Most advanced are the experiments with ions that are kept by electric fields in a high-vacuum chamber. The ions are lined up like a string of pearls. The vacuum prevents them from colliding with other atoms, which would immediately destroy the delicate entanglements. The ground state and a certain excited state serve as the two levels of the qubits. The writing and reading of the information is done with short laser pulses. Important for any successful arithmetic operation is the controlled interaction between the two qubits. This can also be carried out with laser pulses in a way that the oscillation of the ions in the trap is specifically excited with the kick of a photon. This also excites a neighboring ion, which then interacts again with the first excited ion. The entanglement that occurs in this process brings about exactly the intended arithmetic operation.

Another so far quite successful system is the nuclear magnetic resonance in molecules. In this case, the qubits are the different tuning possibilities of the magnetic dipole moments of the atomic nuclei. The qubits are addressed with suitable sequences of radio-frequency pulses. The interaction between two qubits is mediated by the interaction of the nuclei over the electrons of the molecule. Depending on the tuning of the spin, the interaction can be switched on and off and in this way controlled entanglements are formed and calculations are carried out. One problem here is how to address single molecules. Normally the method is used with a very high number of identical molecules and the measurement results correspond to mean values. There is still a discussion going on as to whether this can be seen as a "true" quantum computer.

As a third example, we should mention the quantum computers based on solid states. They are now only in preparation, but soon they are probably going to play a more significant role. Different approaches are based on the Josephson effect or on the so-called quantum dots (artificial atoms). The qubits in an ensemble of superconductors could be addressed in the same way as in conventional computers with transistors. The problem here lies in the extremely short lifetime of the coherence caused by the nearby atoms in a solid. By trying to work at extremely low temperatures one might be able to get around this. Still, many hurdles have to be cleared on the way to the final realization. Decisive progress in this direction should become possible with some help from nanotechnology.

4.7 Fundamental questions and outlook

In quantum computers, domains of the quantum world are understood, which have not been accessible with experiments up to now. This leads us to new basic questions: what kind of quantum algorithms can make best use of the capacities of quantum computers? To what extent can very large quantum systems be selectively influenced "from outside" with control mechanisms that are always classical in the end? When does the classical control of a quantum computer get so complex that its efficiency is reduced? We have to face these questions. An interdisciplinary field of research with basic questions and a promising potential for applications is opening up here at the beginning of the new century.

5

Entangled quantum systems: from wave–particle duality to single-photon sources of light

Gerhard Rempe

Quantum physics is the foundation of modern physics. Even though it was originally developed to explain the phenomena of the atomic domain, spectacular experiments of recent years and new theoretical insights led to an extension of its range of validity to larger objects with higher complexity. This not only led to a fundamentally new assessment of quantum physics, it also made new applications in the information and communication sciences possible. In fact, more than 100 years after its discovery, it has not lost any of its innovative potential. Therefore, quantum physics must be regarded as the basis for all future developments in modern physics.

The growing interest in quantum physics is not at last due to the spectacular progress in the art of performing experiments, which allow the preparation of systems with new kinds of correlations. We refer here to compound systems, for which a measurement at a randomly chosen component defines unambiguously the result of a measurement at another component, even though each single measurement by itself gives an unpredictable result. In other words, the properties of the single components are not necessarily unambiguously defined, but after a measurement of one component, the property of the other component is immediately fixed. The state of the entire system in this case cannot be completely determined by separate measurements of the single components. Instead, the correlations between the components of the system must be taken into consideration for an unambiguous characterization. Schrödinger coined the term

Entangled World, Jürgen Audretsch
Copyright © 2005 WILEY-VCH Verlag GmbH & Co. KGaA, Weinheim
ISBN: 3-527-40470-8

entanglement for such correlations.[1] This term has a very fundamental meaning in quantum physics.

Correlations between quantum objects are not basically new. They occur whenever the individual components of a system interact with each other. New, however, is the fact that more and more refined experimental methods make it possible to assemble quantum objects into a larger system and to specifically entangle them by means of a controlled interaction. These artificially created correlations between the components of a system with possibly macroscopic dimensions are of outstanding relevance, in particular for the information sciences. The research field of quantum optics takes a leading role in these investigations, since one has learned how to control the interaction between quantum-mechanical objects with great precision. This will be discussed further for two selected examples in the following.

The first example is about the wave nature of matter. The wave nature explains experimentally observable effects like interference that occurs when a quantum object moves from one place to another simultaneously along different indistinguishable paths. The wave nature disappears when the single paths are distinguishable. In this case the particle nature of the object shows up. In order to determine the quantum nature quantitatively, the path of the object has to be measured — an everlasting challenge, especially for experimentalists. For this purpose, a new interferometer for atoms was recently developed, in which information about the path of the atom is stored in an internal state of the atom. With this experiment, one can study the quantum-mechanical origin of the wave–particle duality, and moreover carry out a quantitative test of this fundamental principle. From the interpretation of the measurements, it turns out that correlations between different degrees of freedom of the atom are extremely important.

The possibility to form correlations between quantum systems by means of a controlled interaction opens up entirely new perspectives for applications. A particularly far-reaching application potential from the present perspective has the system that consists of an atom and a photon, which will also be considered. Thanks to the spectacular advances in the research fields of atom and light optics, it is now possible to perform experiments with single atoms or photons. In particular,

1) Compare to Section 2.2.

it became feasible to make the coupling between light and matter so strong that a bound state forms between a single atom and a photon. The strong coupling between the atom and the photon in this system can be used to realize a new type of light source, which emits a single photon in a given fixed direction at the push of a button. In this way, in contrast to an attenuated laser beam, a light flash that consists of exactly one photon can be generated. In addition, the information that is stored in the internal energy states of the atom could be transferred to the photon and transmitted to a distant receiver. The system is thus an ideal interface between light and matter that could link together two important elements in a future network of quantum processors.

In the following, we will only introduce the basic ideas behind the experiments. We will therefore dispense with any detailed description of the apparatuses. However, elementary concepts of quantum theory are used for the description of the experiments and the reader should be familiar with these concepts.

5.1 The wave–particle duality

The ability of quantum-mechanical objects to behave either as a wave or as a particle under different experimental conditions is called the wave–particle duality. This is usually illustrated with the double-slit experiment known from optics, which proves the wave nature of light. It is rather surprising that an interference pattern is also formed behind the double slit when the experiment is carried out with massive objects like atoms. This observation can only lead to the interpretation that an atom can also behave like a wave. However, the idea that a massive object could simultaneously move through two spatially separated slits is extremely nonillustrative from the point of view of classical physics. This is why this idea was repeatedly reconsidered in the past: what happens when the object is observed while it passes through the double-slit arrangement? Gedanken experiments, ingeniously worked out, were in the center of these considerations, because at the time when quantum physics was developed true laboratory experiments to clarify the situation could not be made, for technical reasons. The conclusion was that the interference pattern would inevitably get destroyed when the path of the object is known. Namely, as soon as one knows through which one of the two slits the object passes, it behaves as a classical particle that moves on a well-defined path through the apparatus. In this case, the particle nature of the object becomes evident, which prohibits the formation of any interference pattern. Obviously the wave properties and the particle properties of a quantum object exclude each other. The wave–particle duality expresses the fact that it is impossible to observe the wave properties and the particle properties at the same time.

The standard illustrative explanation for the loss of the interference pattern in a so-called which-path experiment is based on Heisenberg's uncertainty relation for position and momentum. This relation says that it is impossible to measure accurately position and momentum of a quantum object at the same time. The close connection between the uncertainty relation and the wave–particle duality was illustrated repeatedly anew in the gedanken experiments mentioned above. A famous example is Einstein's suggestion for the measurement of the path of the object through the double-slit arrangement by means of the recoil on the apparatus. In this case, Bohr was able to show that the uncertainty relation used for position and momentum of the arrangement can explain the loss of the interference fringes.[2,3] A different, but not less famous example comes from Feynman, who picked up a gedanken experiment from Heisenberg and suggested to determine the path of the object with a light microscope (Feynman et al. (1965)). The essential ideas of his suggestion will be reviewed in the following, because they are relevant for the experiment discussed further below.

Feynman considered the arrangement sketched in Fig. 5.1 with two permeable slits. Massive objects, like atoms, move from a distant source towards the double slit. A screen is placed at a large distance behind the double slit. The spatial distribution of the atoms that passed through the arrangement is measured on this screen. It turns out that a pattern of bright and dark fringes is formed. In the dark areas, no atoms are detected. This fringe pattern proves the wave nature of atoms. It materializes because the waves passing through both slits superimpose constructively or destructively depending on their relative phase. In contrast to the interference effects of macroscopic waves, it is irrelevant whether the apparatus contains many atoms. This interference is rather due to the single atoms that move simul-

2) Compare to Chapter 3.
3) See also Section 1.5.

taneously through both slits. The interference pattern emerges after averaging over many atoms that have passed one by one through the double-slit arrangement.

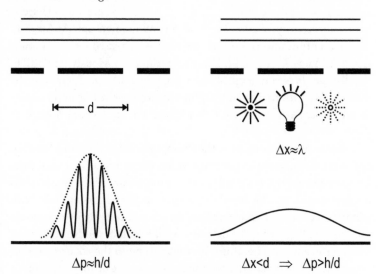

Fig. 5.1 Gedanken experiment from Feynman for the illustration of the wave–particle duality (Feynman et al. (1965)). By localizing the atoms behind the double slit, the uncertainty of the momentum increases. The reason for this is, on the one hand, the position–momentum uncertainty relation, on the other hand the light force that acts during the light scattering. Due to this, the interference fringes are washed out and their envelope is broadened.

Obviously the separation between the interference fringes grows with the distance to the screen. In order to obtain a measure for the fringe separation that is independent of this distance, one should obviously look at the angular separation between neighboring maxima and minima. This angular separation is independent of the distance to the screen, and it is determined by the momentum of the atoms. Atoms that hit the screen at neighboring maxima or minima are distinct by a momentum of the absolute value h/d, where h is Planck's constant (also named Planck's "Wirkungsquantum" in German) and d is the slit distance. The predicted interference pattern was confirmed in many laboratories worldwide during the last years.

Feynman suggested that the atoms should be irradiated with light, in order to observe their path when they pass through the double-slit arrangement, and to detect their position with a microscope. Since the resolution of a microscope is defined by the light wavelength λ, this has to be smaller than the slit distance d. Therefore, each single atom can be localized with an accuracy $\Delta x \approx \lambda \leq d$. Due to Heisenberg's uncertainty relation for position and momentum, $\Delta x \Delta p \geq h$, the momentum uncertainty of the atom is increased by the amount $\Delta p \geq h/\Delta x \approx h/\lambda \geq h/d$. This uncertainty is obviously larger than the distance between neighboring fringes and, hence, washes out the interference pattern. In addition, the envelope of the interference pattern is broadened — an effect that has so far rarely been noticed in the literature.

The microscopic origin for the disappearance of the fringe pattern is the recoil the atom experiences when scattering a light quantum. This scattering process disturbs the momentum of the atom in an uncontrolled way. The recoil disappears only when using light with a long wavelength. However, the resolution of the microscope would then be insufficient for the detection of the path of the atom through the interferometer. Therefore, Feynman came to the conclusion that an experimental apparatus for the measurement of the path cannot be careful enough, so that any uncontrolled disturbance of the quantum object could be excluded. It was thought for a long time that the disappearance of the interference pattern due to a path measurement is always due to Heisenberg's position–momentum uncertainty relation. Obviously this result was not questioned by the nice interference experiments with neutrons (H. Rauch in Bergmann-Schäfer (1993)).

The situation changed only after Scully, Englert and Walther suggested a new double-slit experiment, in which the loss of the interference pattern cannot be explained by Heisenberg's position–momentum uncertainty relation (Scully et al. (1991)). Instead, it was argued that correlations between the path of the object through the interferometer and the detector used for the determination of this path are responsible for the loss of the ability to interfere.[4] In the beginning, there was a controversial discussion about this conclusion and several experiments were initiated, which could meanwhile clarify the situation by using the latest experimental methods. In the following, we highlight one of these experiments, which illustrates

4) See also Section 2.3

the present status of the discussion in a particularly nice way. The experiment uses an atom interferometer, in which a microwave field stores information about the path of the atom in the internal state of the atom. It differs from Feynman's light microscope, besides other aspects, by the fact that the atoms are exposed to microwave radiation of an extremely long wavelength, which causes no (more precisely: a negligibly small) change of the momentum of the atom. The new experiment demonstrates in an impressive way that information about the path of a quantum object through an interferometer can be gained also without the otherwise normal mechanical perturbations.

Fig. 5.2 shows a scheme of the atom interferometer employed (Dürr et al. (1998a)). The Bragg diffraction of atoms at standing light waves is used for splitting and recombining an atomic beam. This process was investigated in detail over recent years and it corresponds to the diffraction of X-rays at a periodic crystal as known from optics — with the difference that the roles of light and matter are exchanged, since matter waves are now diffracted from a grating of light. The first standing wave serves as a symmetric beam splitter that splits the incoming atom beam, A, in exactly two beams, B and C, with the same flux. These two beams are split once more into two components with a second standing wave, in the same way. We thus have two beams, D and E, which propagate to the left and two beams, F and G that propagate to the right. The two beams of each pair overlap in the far field and generate two interference patterns, one to the right and one to the left that are spatially separated from each other. The two interference patterns are complementary to each other in the sense that the maxima of the left pattern correspond to the minima of the right pattern — like the exits of an interferometer for light cannot be both bright or dark at the same time. For the detection of the interference pattern, the atoms are irradiated with laser light and the scattered photons are detected with spatial resolution. Figure 5.3(a) shows an experimentally observed interference pattern. The distance of the fringes is given by the spatial separation of the beams B and C at the second standing wave (about 1 μm in the experiment). The envelope, drawn as a dashed line, is determined by the width of the atom beam in the plane of detection, and can be controlled using diaphragms (not drawn in Fig. 5.2). The last diaphragm is located directly in front of the interferometer, and it has a free aperture of 450 μm.

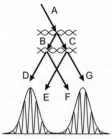

Fig. 5.2 Scheme of the atom interferometer. The atomic beam is split several times by Bragg–diffraction at standing light waves. The beams B, D and E go to the left, the beams A, C, F and G go to the right. The beams D and E (as well as F and G) superimpose in the far field. An interference pattern is formed, with a spatial periodicity — similar to the double slit of light optics — that depends on how far apart the two partial beams B and C are in the second standing wave. This distance can be tuned by adjusting the separation between the two standing waves. Laser-cooled atoms falling through the arrangement under the influence of gravity with a velocity of 2 m/s are used in the experiment. The advantage of the cold atoms is their large wavelength, which leads to an interference pattern with a large fringe distance.

Slow rubidium atoms of the isotope 85, which move in free fall with a velocity of just a few meters per second through a vacuum chamber, are used in the experiment. The interferometer is set up inside the chamber. Slow atoms have the advantage of a large wavelength. In this way, the experimental effort for the observation of the interference pattern is reduced. The two standing waves are realized by light pulses, which are irradiated one after the other. The use of rubidium atoms has a simple technical advantage: the atoms have one valence electron in the outer shell, of which the spin, according to the laws of atomic physics, can be parallel or antiparallel to the nuclear spin. Corresponding to the two alignments, the ground state of the atom splits into two states, $|2\rangle$ and $|3\rangle$, with a slightly different energy. The two states are characterized by the angular momentum quantum numbers $F = 2$ and $F = 3$, respectively. These states have a very long lifetime, and are ideally suited for storing information about the path of the atom. For this purpose a trick is applied, which is based on the fact that atoms in the state $|2\rangle$ suffer a phase shift of 180° when they leave the standing wave in the transmitted beam.

In other words, the sign of the wave function will be inverted for these atoms. In contrast, transmitted atoms in the state $|3\rangle$ or deflected atoms do not experience any phase shift: for these atoms, the wave function remains unchanged. In the following paragraph, this trick will briefly be explained with an analogy known from optics. The reader who is not interested in the details can skip this paragraph.

The trick employs the fact that the frequency of the standing light wave can be tuned in the experiment in such a way that it lies exactly in the middle between the transition frequencies from the state $|2\rangle$ or $|3\rangle$, respectively, to some other (not specified) electronically excited state $|e\rangle$ of the atom. An atom in the energetically lower-lying state $|2\rangle$ then sees a light field with a frequency that is too small to induce a transition from the state $|2\rangle$ to the state $|e\rangle$. However, the red shifted light field induces an oscillating dipole moment in the atom, which oscillates in phase with the driving field. Due to the interaction of the dipole moment with the electric field of the light, a force acts on the atom and accelerates it in the direction of higher intensity. Therefore, the velocity, and with it the wavelength of the atom, is changed by this interaction with the light field. In the present case, atoms in the state $|2\rangle$ get faster, therefore their wavelength becomes shorter. This can be interpreted in such a way that atoms in the state $|2\rangle$ are effectively moving in a medium with a higher refractive index. In analogy to light optics, it can therefore be expected that the matter-wave experiences a phase shift of $180°$ at the reflection from an optically denser medium. This is obviously the case when the atom is in the state $|2\rangle$. A more precise analysis shows that this phase shift occurs when the atom in the state $|2\rangle$ is transmitted. It can be shown in an analogous way that the atomic state $|3\rangle$ experiences no phase shift.

This phase shift for an atom in the state $|2\rangle$ can be transformed into an experimentally measurable population difference between the states $|2\rangle$ and $|3\rangle$. For this purpose, each atom before entering the interferometer is prepared in the state

$$|\psi_A\rangle \otimes |2\rangle \qquad (5.1)$$

where $|\psi_A\rangle$ characterizes the motion of the center of mass of the incoming atom. The symbol $\otimes$ indicates that the state of the entire system can be written as a product of two states, which belong to different degrees of freedom. The atom is afterwards exposed to two microwave fields with frequencies that correspond to the transition frequency between the states $|2\rangle$ and $|3\rangle$. The first field is irradiated in front of the first beam splitter and the second directly behind it. Both fields have a so-called "pulse area" of $\pi/2$. This notation goes back to nuclear spin resonance spectroscopy and it means that the spin of the electron is rotated by $90°$. In other words, the first microwave field transfers the state $|2\rangle$ into the superposition state $|3\rangle + |2\rangle$. Altogether, the state

$$|\psi_A\rangle \otimes (|3\rangle + |2\rangle) \tag{5.2}$$

is formed, where a normalization factor is ignored.

The optical standing wave splits the atomic beam and changes the state vector of the atom to

$$|\psi_B\rangle \otimes (|3\rangle + |2\rangle) + |\psi_C\rangle \otimes (|3\rangle - |2\rangle) \tag{5.3}$$

where the negative sign is due to the phase shift of $180°$, as explained above. The two states $|\psi_B\rangle$ and $|\psi_C\rangle$ denote the state vectors of the motion of the center of mass of the reflected and the transmitted beam (B and C in Fig. 5.2). The second microwave field acts on both beams (B and C) and changes the state vector of the atom to

$$|\psi_B\rangle \otimes |3\rangle - |\psi_C\rangle \otimes |2\rangle \tag{5.4}$$

In the course of this process, the superposition state $|3\rangle + |2\rangle$ is transferred to the state $|3\rangle$. Here, both microwave fields act together, hence like a field with the double "pulse area" π, and the spin is turned by $180°$. In the transmitted beam, the atomic state $|3\rangle - |2\rangle$ is transferred back to the state $|2\rangle$. In this case, the phase shift of $180°$ effectively has the consequence that the two microwave fields cancel each other.

In contrast to Eq. (5.1), Eq. (5.4) shows that the total state of the atom can no longer be written as a product of separate states, individually representing the internal and external degrees of freedom. The state is therefore called nonseparable. It is a strange property of this new state, that a measurement at the internal state of the atom, for example, gives the results $|2\rangle$ or $|3\rangle$ with the same probability. The atom can thus equally be in both states. The same is true for the state of motion: the atom follows with equal probability the paths B and C. However, if a measurement is carried out at one of the two degrees of freedom, the state of the other degree of freedom is unambiguously determined. This correlation of different degrees of freedom is called entanglement. A consequence of entanglement is the fact that, with a measurement of the internal state of the atom, one can find out which path the atom took through the interferometer. For example, when the atom is found in the state $|3\rangle$ in a measurement, it follows immediately that the atom was reflected at the first beam splitter and therefore took the path B. As an interesting fact, this measurement can be carried out later at any time. In summary, it can be concluded that the entanglement is the central point for the storage of the path information in internal states of the atom. Experimentally this is achieved by operating the first beam splitter between two suitable microwave fields.

Figure 5.3(b) shows the distribution of the atoms when the information about the path is stored as described. The experiment clearly proves that the interference fringes disappear. However, it is interesting that the stored information about the path has not been read out, because the detection of the atoms was not made in a state-specific way. Obviously the interference pattern already disappears when the information about the path had been stored and could have been read in principle.

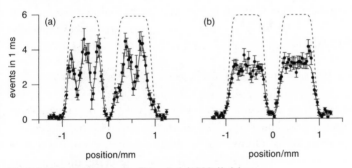

Fig. 5.3 Experimental results (Dürr et al. (1998a)): (a) an interference pattern recorded with ultra cold atoms. Remarkable is the unusually large fringe distance of about 1 mm, which is due to the long wavelength of the slow atoms. (b) Storage of information about the path of the atom destroys the interference fringes; the envelope, however, is not broadened. This proves that no forces occur, which could be made responsible for the vanishing of the fringe pattern.

A careful analysis of the mechanical effects that could occur during the storage of the information about the path shows that Heisenberg's uncertainty relation for position and momentum cannot explain the observed loss of the interference fringes (Dürr et al. (1998a)). In particular, the momentum transfer that occurs with the absorption of a long-wavelength microwave photon is negligibly small because the radiation has a wavelength of about 10 cm. Experimental evidence for the failure of the uncertainty relation is the fact that with the storage of the information about the path, the envelope of the interference pattern is not broadened. The loss of the interference can therefore not be explained by fringes that are just smeared out, i.e. by an accidental superposition of several interference patterns with maxima and minima shifted against each other by at least half a period.

Now we investigate why the interference fringes are lost. For this purpose, we have to look at the state vector of the atom after the passage through the whole interferometer. The state vector after the interaction with the two microwave fields and the first beam splitter has already been discussed above. The second beam splitter transfers this state to

$$|\psi_D\rangle \otimes |3\rangle - |\psi_E\rangle \otimes |2\rangle + |\psi_F\rangle \otimes |3\rangle + |\psi_G\rangle \otimes |2\rangle \tag{5.5}$$

The sign of $|\psi_G\rangle$ is positive, because a phase shift of $180°$ occurs again at the transmission of atoms in the state $|2\rangle$ through the second beam splitter. In the far field, the distribution of the atoms that move, for example, to the left can be determined from the expression

$$|\psi_D(z)|^2 + |\psi_E(z)|^2 - \psi_D(z)^*\psi_E(z)\langle 3|2\rangle - \psi_E(z)^*\psi_D(z)\langle 2|3\rangle \tag{5.6}$$

Note that here the contributions of the position-dependent wave functions $\psi_F(z)$ and $\psi_G(z)$ disappear. The first two terms describe the distribution of atoms, which results when the atoms take the paths D or E towards the detection zone. Therefore, they give the mean intensity under the envelope. Interference comes about only through the last two terms, because they contain the product of the wave functions $\psi_D(z)$ and $\psi_E(z)$. These two terms vanish, though, because $\langle 3|2\rangle = \langle 2|3\rangle = 0$. Physically, the vanishing of this quantity means that both atomic states are perfectly distinguishable. The fact that the internal states can be distinguished guarantees that the two paths can also be distinguished.

Now the question arises, if the disappearance of the interference pattern is not simply based on the possibility that atoms in the states $|2\rangle$ or $|3\rangle$ represent different objects, which cannot interfere with each other for principle reasons. The situation turns out to be slightly more complicated, because the information about the path of the atom can quite easily be erased: when, for example, atoms in the state $|2\rangle +$ $|3\rangle$ are detected — which can easily be achieved experimentally by irradiating an additional microwave field with the "pulse area" $\pi/2$ — this results in the state

$$|\psi_D\rangle - |\psi_E\rangle + |\psi_F\rangle + |\psi_G\rangle \qquad (5.7)$$

after the second beam splitter.

This state describes four waves of which two, namely D and E, move to the left. The other two waves, F and G, move to the right. A similar situation was already described in Fig. 5.1. Therefore, the pairwise superimposed waves interfere with each other again — and in fact with full contrast. It follows that the interference pattern can be restored through a suitable measurement in a subensemble of the atoms. A similar result is obtained for the case that atoms are detected in the state $|2\rangle - |3\rangle$, orthogonal to $|2\rangle + |3\rangle$. The state after the second beam splitter is then

$$-|\psi_D\rangle - |\psi_E\rangle - |\psi_F\rangle + |\psi_G\rangle \qquad (5.8)$$

This state differs from the state described in Eq. (5.7) just by an exchange of two signs. Therefore, four states result also in this case, but with a different phase, reflected in the two negative signs. This leads to the situation that maxima and minima of the interference patterns, characterized by Eqs. (5.7) and (5.8), are exchanged. The sum over the two distributions shows in consequence no interference. This was expected, because all atoms together produce no interference pattern. It follows that no fundamentally different objects are generated with the storage of the path information. Since the interference can be restored, the coherence in particular is not lost.

In summary, the entanglement, which is necessary for storing the information about the path of the atom, is also responsible for the loss of the interference. In other words, correlations between the which-path detector and the atomic motion destroy the interference fringes. Correlations of course also exist in Feynman's gedanken experiment. There, however, the vanishing of the interference pattern can be explained with the far more illustrative mechanical effects occurring during the localization. Such an explanation based on Heisenberg's uncertainty relation for position and momentum does not lead to the goal in the case of the discussed experiment with the atomic interferometer.

The correlations between the path of the atom and the internal state are not required to be perfect. In the experiment, particularly the amplitude of the microwave field and with it the degree of correlation can be continuously tuned: when the microwave field rotates the spin of the electron by any angle φ (the "pulse area"), the superposition of the two states $|2\rangle$ and $|3\rangle$ can be varied. In this case, the state of the atom after the second microwave field is:

$$|\psi_B\rangle \otimes (\cos\varphi|2\rangle + \sin\varphi|3\rangle) - |\psi_C\rangle \otimes |2\rangle \qquad (5.9)$$

As a consequence, only an incomplete knowledge about the path of the atom can be gained with a measurement of the internal state of the atom later. It is, in particular, no longer possible to conclude unambiguously, for example, from the measurement result $|2\rangle$ about the path C, as it was possible for the experiment discussed above with $\varphi = \pi/2$. Instead, only probabilities that one or the other path was taken by the atom can be given. In addition, the previously mentioned phenomenon results, namely that these probabilities depend on which of the observables of the atom is measured. When, for example, atoms in the state $|2\rangle + |3\rangle$ are detected, the possibility to unambiguously identify the path B is lost. The consequence then is that the maximum information that can be stored is not necessarily read out when making an observation of the internal state of the atom. In fact, the information — as discussed above — can be irretrievably erased. Therefore, one has to find a suitable observable, for which the information about the path of the atom is at maximum. When this observable is measured, the information obtained in this way is called the distinguishability of the paths D (Englert (1996)). In the case $D = 1$, the paths can be perfectly distinguished and the object shows its particle character. For $D = 0$, no which-path information exists.

When the information about the path is only incompletely stored (or present), the interference pattern does not vanish completely — only its visibility V is reduced. Experimental results are presented in Fig. 5.4 for different values of the microwave amplitude. Figure 5.4(a) corresponds to Fig. 5.3(a) and it shows once more an interference pattern recorded without a microwave field. The difference is just that in Fig. 5.4(a) the two partial beams B and C known from Fig. 5.2 went less far apart. This explains the larger fringe distance compared to Fig. 5.3(a). Figure 5.4(c) corresponds to Fig. 5.3(b) and it demonstrates the vanishing of the interference pattern for a microwave field with a pulse area of $\varphi = \pi/2$. Figure 5.4(b) was recorded with a pulse area of $\varphi = \pi/3$. In comparison with Fig. 5.4(a), the visibility V is significantly reduced. A pulse area of $\varphi = \pi$ was used in Fig. 5.4(d). The consequence is that the atoms are in the internal state $|3\rangle$ when they enter the interferometer. Therefore, the contrast is inverted but the amplitude of the fringe pattern and, hence, the visibility remains at maximum.

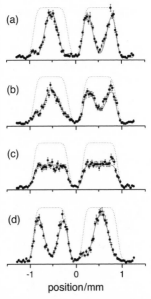

Fig. 5.4 Interference pattern for different amplitudes of the microwave field: (a) no microwave field, hence pulse area $\varphi = 0$, (b) pulse area $\varphi = \pi/3$, (c) pulse area $\varphi = \pi/2$ and (d) pulse area $\varphi = \pi$. The interference vanishes for $\varphi = \pi/2$. In the case of $\varphi = \pi$, the contrast is inverted: maxima and minima are exchanged. The envelope of the interference pattern (dashed line) remains unchanged.

The visibility V and the distinguishability of the paths D are not independent of each other, but they are linked together by the so-called duality relation (Englert (1996))

$$D^2 + V^2 \leq 1 \qquad (5.10)$$

This inequality, which is a quantitative formulation of the wave–particle duality, has been found just recently. It goes beyond the usual interpretation of the wave–particle duality, because intermediate situations obviously exist, in which the object shows both wave and particle properties. In addition, it also makes a quantitative test of this principle, fundamental for quantum physics, possible.

The first experimental tests of the duality relation were carried out recently. The two quantities D and V have been measured independently for different values of φ. The results shown in Fig. 5.5 are recorded with the atom interferometer described above. They agree well with the duality relation. The fact that the observed visibility V and the distinguishability of the paths D do not reach the theoretically expected maximum value of 100 % is due to shortcomings in the experimental setup.

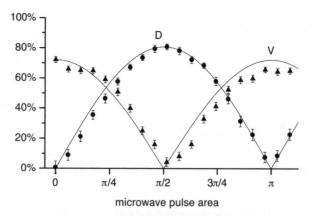

Fig. 5.5 Visibility of the interference pattern V (triangles) and distinguishability of the paths D (dots) as a function of the pulse area φ of the microwave fields used for storing the path information. The full line shows the theoretical prediction (see Dürr et al. (1998b)).

Note that in the experiment with the atom interferometer, the internal state of an atom was entangled with its external state. This concerns always just one single quantum object — the atom. In the following, we introduce a system in which two different quantum objects — an atom and a photon — are coupled with each other. Such a system makes new applications possible, which are based on the fact that the atom and the photon can be measured independently of each other.

5.2 The quantum interface

Internal states of an atom are ideally suitable for the storage of information (Bouwmeester et al. (2000)). This possibility was already used in the atom interferometer discussed above. At present, however, it appears to be not realistic to use the atoms for a transfer of information over a larger distance. The reason for this is the fact that atoms usually move slowly and, in addition, they can easily collide with other atoms. Through collisions they change their direction of motion, so that the receiver is waiting in vane for the arrival of the information. Also, the inner state of the atom can change in the course of the collision and thus the stored information can get lost. Light quanta, which can be transported in glass fibers over long distances rapidly and almost without losses, are much more suitable for the transfer of information. The development of an interface between light and matter, which allows the transfer of information from one system to the other and back, is therefore important. Small resonators with the highest quality factors have a special significance in this context, because inside, atoms at rest can be coupled with photons in a unique way. This should first be explained with a simple consideration.

Imagine an atom in an excited energy level. The atom will not stay excited for long, but as a rule, it will return to the ground state after a short time. The excitation energy is then emitted in the form of a light quantum. This process is called spontaneous emission. Formerly, this decay process was considered to be a fundamental physical property of the atom, which cannot be influenced. Only recently it has been realized that the environment of the atom plays an important role in this process. In the case of an atom between two highly reflective mirrors, for example, the emitted light quantum is always reflected to and fro: The two mirrors form an optical resonator in which the light quantum is stored. When it is stored for long enough, it can be absorbed again from the atom. The atom returns then to the excited state and the process of emission followed by reabsorption starts all over again. For an atom between two mirrors, the otherwise irreversible process of spontaneous emission can become reversible — an extremely interesting situation.

Under which conditions can this periodic energy exchange be realized? An important requirement is a small mirror distance. This can easily be understood, because the energy of the photon is then in a small volume. As a consequence, the interaction energy of the photon with the atom, and thus the rate of exchange of excitation energy between the atom and the light field, is large — an important precondition for the experiments. However, the smaller the resonator, the shorter is the dwell time of a photon inside, because the light quantum reflected to and fro is repeatedly hitting a mirror. At each reflection, the photon has a small probability of being transmitted. This disadvantage of short storage times can only be compensated by using mirrors with the highest possible reflectivity in the experiments. Therefore, a very small resonator with very good mirrors is needed.

An enormous development in mirror production took place in recent years. Nowadays, it is possible to produce mirrors with remaining losses due to absorption and transmission of only 0.0001 %. This small value is at least a factor of ten thousand better then for the mirrors made of aluminum or silver and it can only be achieved with the latest technology. Up to 40 dielectric (not absorbing) layers with different refractive indices are deposited on a glass substrate. Before that, the substrate is polished so smoothly that the remaining roughness is lower than the diameter of an atom. This guarantees very plane layers with a minimum of scattering losses. Every interface between the single layers reflects a part of the light. The layer thicknesses are chosen in such a way that the reflected beams amplify each other by interference. Fantastic reflectivity values of up to 99.9999 % can thus be achieved. With the best mirrors available at present, light can be reflected to and fro a few hundred thousand times. These mirrors from the newest technology are of immense importance for the atom–photon experiments.

As a note, the periodic energy exchange between two coupled systems discussed above can be found in many situations. It occurs, for example, already in the simple case of two mechanical pendulums coupled with each other by a spring. It is well known that the pendulum pushed once transfers its kinetic energy to the other, initially resting pendulum: The two pendulums are alternately moving to and fro. An oscillatory behavior invariant in time exists only when both pendulums are oscillating with the same amplitude, either with the same or the opposite phase. These two so-called characteristic oscillations are shown in Fig. 5.6. For the oscillation in Fig. 5.6(a), the spring is periodically stretched and compressed. The force, which is necessary for this process, leads to an additional acceleration of the pendulums, so that the oscillation mode (a) has a higher frequency than the oscillation mode (b). The frequency difference between the two oscillation modes determines the frequency of the periodic energy exchange: This goes faster the stronger the two pendulums are coupled over the spring.

(a)
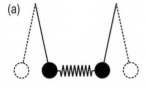

(b)
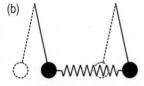

Fig. 5.6 Model that illustrates the oscillation modes of a coupled system of two mechanical pendulums. In (a) the pendulums are oscillating in the opposite direction, in (b) in the same direction. Mode (a) has a higher oscillation frequency than mode (b).

A consideration analogous to the case of the two pendulums presents itself now for the strongly coupled atom-photon system, where the atom and the light field take the part of the two pendulums. The state $|e\rangle \otimes |0\rangle$ corresponds then to the motion of one pendulum, while the state $|g\rangle \otimes |1\rangle$ corresponds to the motion of the other pendulum. In this case, the state $|e\rangle$ ($|g\rangle$) is the excited state (ground state) of the atom and $|0\rangle$ ($|1\rangle$) is the state of the light field with none (one) photon in the resonator. Two eigenstates of the coupled system exist, correspondingly. One of the states, $|-\rangle$, has an energy that is lower by the amount E than for the uncoupled system. Formally, it is characterized by the term

$$|e\rangle \otimes |0\rangle - |g\rangle \otimes |1\rangle \qquad (5.11)$$

The other eigenstate, $|+\rangle$, is given by the term

$$|e\rangle \otimes |0\rangle + |g\rangle \otimes |1\rangle \qquad (5.12)$$

and it has an energy that is higher by the same amount E. The energy separation of the two states, $2E$, is proportional to the frequency of the energy exchange between the atom and the light field.

This now allows us to study in more detail the periodic energy exchange between an atom and a light field in a resonator that was already introduced above. For this purpose, an atom in the excited state is brought into an empty resonator at the time $t = 0$. The whole system is then described by the state vector

$$|\psi(0)\rangle = |e\rangle \otimes |0\rangle \qquad (5.13)$$

By using the Eqs. (5.11) and (5.12), this initial state can also be written as the sum over the energy eigenstates $|+\rangle$ and $|-\rangle$. The time evolution of the state vector $|\psi(t)\rangle$ can then be calculated with the term

$$|\psi(t)\rangle = \exp(-igt)|+\rangle + \exp(igt)|-\rangle \qquad (5.14)$$

following the usual formalism of quantum physics, where a normalization factor was ignored and $\hbar g = E$ was set ($\hbar = h/2\pi$). From this, it can be directly deduced that the atom is in the ground state $|g\rangle$ after half a period ($gt = \pi/2$), and that then a photon is in the resonator. The whole state is then given by $|\psi(gt = \pi/2)\rangle = -i|g\rangle \otimes |1\rangle$. After a full oscillation period ($gt = \pi$), the original state is restored, apart from a change in the sign: $|\psi(gt = \pi)\rangle = -|e\rangle \otimes |0\rangle$. But what happens after a quarter of a period ($gt = \pi/4$)? Obviously the whole system is then described by the entangled state

$$|\psi(gt = \pi/4)\rangle = |e\rangle \otimes |0\rangle - i|g\rangle \otimes |1\rangle \qquad (5.15)$$

One property of this state is, for example, that a measurement at the light field gives the photon-number states $|0\rangle$ and $|1\rangle$ with the same probability. So, the light field is not in a state with a fixed photon number. The analogous is true for the internal state of the atom.

However, even in this case, through the measurement at one sub-system information about the other subsystem can be obtained — similar to the atom interferometer. For example, the photon number in the resonator can be determined with a suitable measurement at the atom.

The transfer of the locally present entanglement between the atom and the light field to a different, extended system should in principle be possible. The atom could leave the resonator after a quarter of a period and fly to a different, possibly far-distant resonator. When the atom is in the state $|e\rangle$, it could emit its excitation energy there and create a state with a photon in the second resonator. An atom in the ground state $|g\rangle$ has no effect, so that the second resonator would remain empty in this case. Then the following state

$$|1_2\rangle \otimes |0_1\rangle + |0_2\rangle \otimes |1_1\rangle \qquad (5.16)$$

results, where the indices 1 and 2 mark the two resonators. In this way, the light fields of two spatially separated resonators can, in principle, get entangled with each other and a quantum system with macroscopic dimensions can be created. However, such an experiment has so far not been realized. Besides the disadvantages mentioned above, the excited atom flying from one resonator to the other can spontaneously emit a light quantum in an uncontrollable direction, a fact that has to be taken into account. The photon that actually should have been placed in the second resonator would then get lost on its way. For practical applications it is therefore far more interesting to follow a different route and couple the photon out from the first resonator and send it into the second resonator. There, a second atom could absorb this photon. The state

$$|e_1\rangle \otimes |g_2\rangle - |g_1\rangle \otimes |e_2\rangle \qquad (5.17)$$

can then be formed, where the indices 1 and 2 now mark the two atoms. In this way, it is in principle possible to create a specific entanglement of the internal degrees of freedom of two spatially separated atoms — a perspective that looks extremely enticing from the point of view of information technology. The condition is just that the transfer of the information present in one atom can be achieved without losses from one resonator to the other. On the side of the sender, a light source is needed that emits a photon in a well-defined direction, namely to the other resonator. However, this experiment has not been carried out yet. Only recently a single-photon light source meeting the high requirements of information technology was successfully developed. This system is now introduced.

5.3 The single-photon light source

One of the greatest challenges for the development of an efficient single-photon light source is the emission of the light quantum in a well-defined direction. This is particularly so for light sources based on spontaneous emission, because here the photon is emitted in an unpredictable direction for reasons of principle. This disadvantage can not be avoided with an atom in a resonator, since the mirrors typically cover only a very small solid angle. This holds in particular when the distance of the mirrors is very small so that they cannot have a strong curvature. In this case, the atom can emit its excitation energy also in a direction perpendicular to the axis of the resonator. This process cannot be completely suppressed even when the coupling of the atom to the resonator is very strong. In order to avoid completely any spontaneous emission, the atom must not be in the excited state at any time. But is it possible at all then to emit a photon into the resonator? And when it is, where does the energy come from?

In order to explain the mode of operation of the new light source, we should first discuss a simple mechanical model. This is illustrated in Fig. 5.7 and it consists of three pendulums coupled with each other. The springs mounted in between the pendulums have a different strength. In other words, the pendulums are coupled with each other differently.

In Fig. 5.7(a), the left spring for example is very soft (weak coupling), and the right one, in contrast, is very hard (strong coupling). When, in such a system, the left pendulum is brought into motion, the other two pendulums remain at rest. This has a simple cause: the two right pendulums will only move under the action of a force. In spite of the strong amplitude of the left, weak spring, the force acting on the pendulum in the middle is relatively weak. It will therefore be compensated with some very small amplitude of the right, hard spring. This is why the two other pendulums are not moving. It is, nevertheless, still possible to transfer the motion of the left pendu-

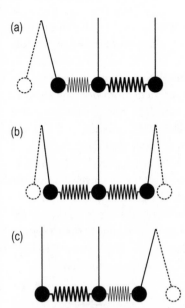

Fig. 5.7 Model that illustrates a slow energy transfer between mechanical oscillators. Three pendulums are coupled with each other by two springs of different strength. The thin (thick) springs are soft (hard) and symbolize a weak (strong) coupling. With a slow change of the spring constants, the initial motion of the left pendulum (a) can be transferred to the right pendulum (c). The pendulum in the middle does not move during this process.

lum to the right pendulums. For this purpose, one only has to invert the coupling ratio, slowly. This is illustrated in Fig. 5.7(c), where the left, originally soft spring is now hard. Correspondingly the right, originally hard spring now becomes soft. For reasons of symmetry, the right pendulum is oscillating while the left pendulum is at rest. In contrast to the case shown in Fig. 5.7(a), though, the right pendulum has to move now to the right in order to stretch the spring. This means that the right pendulum is now oscillating with the opposite phase compared to the left pendulum. This has an important consequence for the interesting intermediate situation (Fig. 5.7(b)) in which the two springs have the same strength. Here, the two outer pendulums are already oscillating out of phase, but with the same amplitude. Therefore, the resulting force on the middle pendulum vanishes, so that it again remains at rest. In fact, the middle pendulum never gets excited, at any moment in time. The only requirement to achieve this condition is that the coupling strength is changed sufficiently slowly.

Similar to the case of the mechanical pendulums, the energy can also be transferred in the optical experiment from one light field to another without exciting the atom, which in this process is needed for the coupling of the two light fields. This situation is illustrated in Fig. 5.8, where a three-level atom is coupled to the light fields of a pump laser and a resonator. The three energy levels take the role of the three pendulums and the two light fields correspond to the two springs of the mechanical model. The laser frequency is close to the frequency of the atomic transition from the state $|i\rangle$ to the excited state $|e\rangle$, the frequency of the resonator corresponds to the transition frequency from the state $|f\rangle$ to the state $|e\rangle$. The states $|i\rangle$ and $|f\rangle$ have a long lifetime, and the state $|e\rangle$ can decay spontaneously.

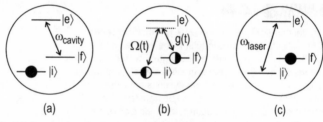

(a)   (b)   (c)

Fig. 5.8 Generation of a single photon by slowly switching off the coupling constant between the atom and the resonator, $g(t)$, and simultaneously switching on the coupling to a laser, $\Omega(t)$. The resonator couples the states $|e\rangle$ and $|f\rangle$, the laser couples the states $|e\rangle$ and $|i\rangle$. The atom is initially in $|i\rangle$, then in a superposition of $|i\rangle$ and $|f\rangle$ and in the end in $|f\rangle$. In the course of this process, a photon is emitted into the initially empty resonator, while the atom is never in the excited state $|e\rangle$.

The initial situation is illustrated in Fig. 5.8(a): the atom is in the state $|i\rangle$, the pump laser is switched off and the coupling to the resonator, characterized by the coupling strength g, is at maximum. The system is therefore described by the initial state

$$|\psi(0)\rangle = |i\rangle \otimes |0\rangle \tag{5.18}$$

where the state $|0\rangle$ marks the empty resonator. When the coupling g to the resonator is now slowly turned off and, simultaneously, the coupling Ω to the pump laser is turned on, the atom is transferred to the state $|f\rangle$. In this process, exactly one photon is generated inside the resonator, while one light quantum is absorbed from the laser beam. The final state is therefore

$$|\psi(\infty)\rangle = -|f\rangle \otimes |1\rangle \qquad (5.19)$$

This state is sketched in Fig. 5.8(c). The negative sign is supposed to indicate the phase shift of 180°, already known from the mechanical model. In analogy to the pendulums, it further follows that the atom is at no moment in time in the excited state $|e\rangle$. The system is rather described by the dark (not radiating) state

$$|\psi(t)\rangle = g(t)|i\rangle \otimes |0\rangle - \Omega(t)|f\rangle \otimes |1\rangle \qquad (5.20)$$

at any time (see Fig. 5.8(b)), where the normalization constant was once again ignored. The time dependence of the coupling constants of the atom to the resonator, $g(t)$, and the pump laser, $\Omega(t)$, are explicitly expressed. The excited state $|e\rangle$ does not occur. The atom, instead, is in a coherent superposition of the states $|i\rangle$ and $|f\rangle$. Correspondingly, the resonator is in a superposition of the two states $|0\rangle$ and $|1\rangle$. The negative sign in Eq. (5.20) takes into account that the transition amplitudes from the states $|i\rangle$ and $|f\rangle$ to the excited state $|e\rangle$ interfere destructively.

But how can the two coupling strengths be varied in the desired way? A laser can easily be switched on, but how can a resonator be switched off? At this point, a trick helps as described in the experimental setup sketched in Fig. 5.9. The trick is based on the fact that atoms are moving, and that the light field in a resonator has a limited spatial extension. The light field of the pump laser irradiating from the side is shifted compared to the light field of the resonator sucht that the two light fields are just still overlapping. The direction of the shift is chosen in such a way that the atom is first coupled to the (still empty) resonator before entering the laser beam while leaving the resonator. Only at this moment does the atom emit a photon under the influence of the two couplings. It is remarkable that the intuitively expected sequence of the light fields — at first the excitation of the atom with the pump laser and then the emission into the resonator — is inverted here: we have first the (empty) resonator and then the (intense) pump laser. Only with this antiintuitive arrangement can it be achieved that the atom emits a light quantum without previously being in an excited state.

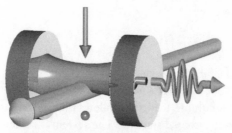

Fig. 5.9 Experimental setup. A rubidium atom, cooled by laser radiation, falls under the influence of gravity at first through a (empty) resonator and then through an intense laser beam. The laser beam is below the axis of the resonator. The resonator consists of two mirrors at a distance of 1 mm. One of the mirrors has a transmission of 0.01 %, the other just 0.0004 %. Due to the small transmission losses, the light can be reflected to and fro in between the mirrors for about 20 000 times. The light quantum that is generated while the atom transits through the system leaves the resonator through the less-reflective mirror, and from there is available for applications.

The described procedure works only in the case of a two-photon resonance, which is defined by imposing the condition that the difference frequency of the two light fields must agree with the transition frequency of the atom from the state $|i\rangle$ to the state $|f\rangle$. Only in this case is the total energy conserved. It is noteworthy that the two light fields do not need to be tuned individually on the two corresponding transitions of the atom. Figure 5.10 shows an experimental result, where the number of the emitted photons is plotted as a function of the frequency of the pump laser for a fixed frequency of the resonator. This frequency is smaller than the atomic transition frequency between the states $|i\rangle$ and $|e\rangle$. Obviously photons are emitted only in a small frequency range around the two-photon resonance. Interestingly, this range is only half the size of the linewidth defined by the lifetime of the atom's excited state. This proves that the excited state of the atom is not populated. This observed linewidth is instead determined by the lifetime of the photons in the resonator.

The new light source is based on the process that single atoms fall one after the other first through the initially empty resonator

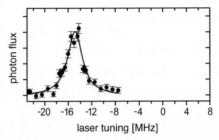

Fig. 5.10 Number of photons emitted from the resonator as a function of the laser frequency at a fixed frequency of the resonator. This is 15 MHz lower than the atomic transition frequency. Light is only emitted within a small frequency interval around the so-called two-photon resonance (see Hennrich et al. (2000)).

and then through the exciting laser beam. The question arises, if the procedure also works for an atom that rests in the resonator. The following consideration shows that this should really be the case: For this method, it is essential that the strongly coupled system consisting of the atom and the resonator is at all times in a (not radiating) dark state. It is also important that the final state $|f\rangle \otimes |1\rangle$ we discussed so far cannot produce any atomic excitation. This was achieved by switching off the coupling constant g. When taking a closer look it becomes clear, though, that this is not necessary at all, because the true final state $|f\rangle \otimes |0\rangle$ is always reached by the time the photon leaves the resonator. This state is dark even in the case that g is kept constant. The reason for this is that the coupling to $|e\rangle \otimes |0\rangle$, which is necessary for a new excitation of the atom, is no longer present. It thus appears to be realistic for a future experiment to transform an atom trapped in a resonator from the final state $|f\rangle$ back to the initial state $|i\rangle$, for example with additional laser beams, and to repeat the emission process. In this way, one might be able to generate a bitstream of single photons with a high efficiency. The described light source, therefore, opens up an interesting perspective for quantum information technology.

Moreover, it should also be possible to transfer an atomic superposition state to the light field (Bouwmeester et al. (2000)). Namely when the system is initially in the state

$$(\alpha|i\rangle + \beta|f\rangle) \otimes |0\rangle \tag{5.21}$$

the interaction described above leads immediately to the state

$$|f\rangle \otimes (\alpha|1\rangle + \beta|0\rangle) \tag{5.22}$$

Obviously the information stored in the atom, characterized by the arbitrarily chosen probability amplitudes α and β, is completely transferred to the light field. This information can be guided to a second, distant system, and be transferred to another atom. For that purpose, only the emission process described above has to be inverted. Therefore the second atom has to be prepared in the state $|f\rangle$ and an additional laser beam has to be switched on, when the light pulse arrives at the second resonator. This condition can easily be met, since for a deterministic light source, the position of the light pulse is known at any time. In this way, the state of an atom can be teleported from one place to another — an extraordinarily tempting possibility of an application for the described single-photon source of light.

Final remarks

Quantum physics is already more than 100 years old, but only in recent years and with the rapid technological progress, more precise investigations of fundamental problems of quantum physics became possible. The ability to control quantum objects with an increasing precision is a particularly important precondition for practical applications. An example of this is the quantum information technology (Bouwmeester et al. (2000)), where information scientists, mathematicians and physicists work together, in order to make good use of the laws of quantum physics in the processing and transmission of information. This field of research is currently experiencing a fast and extraordinarily fruitful development — comparable with the creation of the fundamental mathematical principles of communication and data processing in the 1930s. The basis for this is the core property of entanglement in quantum physics, which makes a completely new route to information processing possible. While mathematicians and information scientists are more interested in the structure of new quantum algorithms, the development of new hardware components is of great relevance for experimental physicists. The atom–photon system is particularly exciting in this context, because in this case the light–matter interaction is very strong — an important prerequisite for the generation of entangled atom–photon states. Quantum technology obviously opens up fascinating possibilities.

The author would like to thank Dr. Stephan Dürr, Markus Hennrich, Dr. Axel Kuhn, Thomas Legero and Thomas Nonn, who performed the experiments at the University of Konstanz and the Max-Planck-Institute for Quantum Optics.

References

- Bergmann-Schäfer (1993), *Lehrbuch der Experimentalphysik*, Walter de Gruyter & Co, Berlin, vol. 3: Optik, chapter 11, 1089–1144.
- D. Bouwmeester, A. Ekert, A. Zeilinger (eds.) (2000), *The Physics of Quantum Information*, Springer-Verlag, Berlin. An introduction into the physical principles of quantum information technology.
- S. Dürr, T. Nonn, G. Rempe (1998a), *Nature* **395**, 33.
- S. Dürr, T. Nonn, G. Rempe (1998b), *Phys. Rev. Lett.* **81**, 5705.
- B.-G. Englert (1996), *Phys. Rev. Lett.* **77**, 2154.
- R. P. Feynman, R. B. Leighton, M. Sands (1965), The Feynman Lectures on Physics, Addison-Wesley, Reading, vol. **III**, chapter 1.
- M. Hennrich, T. Legero, A. Kuhn, G. Rempe (2000), *Phys. Rev. Lett.* **85**, 4872.
- M. O. Scully, B.-G. Englert, H. Walther (1991), *Nature* **351**, 111.

6
Quantum information

Harald Weinfurter

Entanglement, the superposition principle and other elementary effects of quantum mechanics broaden and improve conventional methods of information processing. Quantum cryptography allows for the first time the eavesdropping-proof communication of messages, teleportation makes the transfer of quantum attributes possible, and finally the quantum computer can solve difficult numerical problems within a short time. The first experiments demonstrate these fascinating possibilities and today form the foundation for an information technology of the future.

6.1 On the road to quantum information science

Information transfer and information processing are key technologies of our society. The distribution of the latest news and the fast access to gigantic amounts of data form an important building block of the economic and also political structure. Essential for the development of communication technologies are the enormous progresses in the field of semiconductor technology. A rapid miniaturization of electronic components started with the invention of the transistor and the microchip. This is shown clearest with Moore's law. Already in the year 1965, J. Moore, one of the founders of the company Intel, set the objective that the number of transistors on a chip should double every 18 months. Even though the doubling period was subsequently about two years, there is no other branch of industry in which comparable progress has been achieved over such a long time. Such an avalanche-type (exponential) increase also rarely happens in nature.

Entangled World, Jürgen Audretsch
Copyright © 2005 WILEY-VCH Verlag GmbH & Co. KGaA, Weinheim
ISBN: 3-527-40470-8

Examples are the number of epidemically infected or the growth of a swarm of locusts. However, the growth soon slows down and finally stops, for example when there is not enough food available for all the locusts.

Only the miniaturization of semiconductor components appears to advance without any inhibition and limits — or perhaps not? At the moment, the current between the transistors, i.e. the flow of electrons, can be compared with the flow of water in hoses, in accordance with the laws of classical physics. With continuing miniaturization, fewer and fewer electrons go through thinner and thinner strip conductors. Their propagation sooner or later can only be described by quantum mechanics. When the planned miniaturization is achieved in the future, quantum effects will have to be taken into account, by the latest in 15 years from now. Maybe it will be possible to get around these effects for some time. But in the end, a natural limit is set by the size of the atoms: a conducting path must at least be formed with one chain of atoms. On current microchips the transistors are linked with 300 nm (nanometer, a millionth part of a millimeter) wide strip-conductors; such paths are still at least about 1000 atoms wide. Quite soon the limit will be pushed towards much smaller dimensions.

Is it reasonable, though or anyway necessary, to avoid or to circumvent the quantum effects occurring? In conventional technologies information is encoded by electrical currents or strong light pulses. This makes the digital coding of information possible, which is to a large extent insensitive towards noise and other disturbances. However, when quantum effects occur at the transfer of information, a loss of precision is expected. On the one hand, Heisenberg's uncertainty relation says that certain physical quantities like position and momentum cannot be measured with arbitrary precision at the same time. On the other hand, measurements in the quantum world are in principle of a statistical nature. All this gives rise to the supposition that any quantum effect might cause high noise and a high error rate — this is certainly reason enough to avoid these quantum effects.

The new field of quantum information science shows that basic quantum effects can be used for the formulation of new communication methods and it also shows the way in which this can be done (Bennett (1995), Lo at al. (1998), Bouwmeester et al. (2000)). Exactly by employing the uncertainty relation, quantum cryptography makes a completely eavesdropping-proof transmission of messages possible for the first time. Quantum teleportation allows the transfer of quantum attributes from one particle to another and with this it provides an important building block for a number of additional communication methods. The quantum information science describes not only new possibilities for the transmission, but also for the processing of information. More efficient and faster algorithms were suggested for quantum computers than are possible for conventional computers. When a quantum computer is successfully built, it could perform tasks that appear to be unsolvable at the moment, like for example the prime factorization of large numbers.

This has a certain irony: on the one hand the use of quantum physics makes conventional encoding procedures that rely on the impossibility of an efficient prime factorization for their security completely obsolete and useless all at once. On the other hand, quantum cryptography provides a new possibility for the transmission of messages for which the security is no longer guaranteed by mathematical tricks but instead by the laws of physics.

In the following, it is shown in which way the basic elements of quantum mechanics can be used for new methods of information transmission and information processing. Later, an overview of the actual status of the first experiments for quantum teleportation and for quantum computers, as well as the developments for quantum cryptography is given.

6.2 Basics of quantum information science

The efficiency of the new quantum information procedures is based on an extension of the conventional digital coding of information.[1] The basic unit of information, the bit, is usually expressed by the values "0" and "1". The physical carriers of the information are adapted to the respective technical realization. Current, voltage or light are typically used. For the TTL-logics in microchips for example, the voltage level is 0 V for logic "0" and 5 V for "1". In the case of information transmission over glass fibers, no light is sent for "0" and a short light pulse for "1".

---

1) See also Chapter 7.

What happens now when the light source is attenuated until finally only a single quantum of light — a photon — is transmitted? What should we expect when the switching process of a transistor is already triggered with a single electron?

In quantum information science, quantum objects are used as carriers of information. When two possible settings exist for a particular property, they can be used to express "0" and "1". We are going to use the quantum terminology $|0\rangle$ and $|1\rangle$ in order to distinguish these settings from the classical values of the bit. Figure 6.1 shows a series of possible realizations, for example the linear polarization of a light quantum with the settings $|H\rangle$ (horizontal) and $|V\rangle$ (vertical) for $|0\rangle$ and $|1\rangle$, or the ground and excited state of an atom, respectively.

"0"	"1"	Qubit
$\updownarrow$ $|V\rangle$	$\longleftrightarrow$ $|H\rangle$	photon linear polarisation
$\circlearrowleft$ $|L\rangle$	$\circlearrowright$ $|R\rangle$	photon circular polarisation
$\uparrow$ $|+\frac{1}{2}\hbar\rangle$	$\downarrow$ $|-\frac{1}{2}\hbar\rangle$	electron, neutron, atomic nucleus: spin
$\overline{|g\rangle}$	$\overline{|e\rangle}$	atom, ion: internal states
$|g\rangle$	$|e\rangle$	quantum dots: energy levels
$|a\rangle$ $|a'\rangle$	$|b\rangle$ $|b'\rangle$	particles: modes at the beam splitter

Fig. 6.1 Different possibilities to realize a qubit. With a photon one can use for example linearly or circularly polarized light, with an electron or neutron the spin, with an atom the energy states and with any kind of particle in the two modes in front of or behind a beam splitter.

While there are only two possibilities "0" and "1" allowed for the classical bit, the quantum system can be in any state that results from a superposition of the two basic settings. The general state is then expressed as

$$a_0|0\rangle + a_1|1\rangle$$

In a physical sense, this means that the quantum system has the probability $|a_0|^2$ to be in the state $|0\rangle$ (it has the value "0") and has probability $|a_1|^2$ in the state $|1\rangle$ (value "1"). The value of the bit itself is therefore quantum-mechanically uncertain. An observation will show one of the two values with the given probability as a result. Does this uncertainty not actually go together with a loss of information?

In quantum mechanics one has to make the strict distinction between a superposition and the classical mixture of two possibilities. When we detect, for example, at an ensemble of photons horizontal and vertical polarization with the same probability, it could have been an incoherent mixture without any information content. However, it could have been a coherent superposition as well, for example

$$(|H\rangle + |V\rangle)/\sqrt{2} = |45°\rangle$$

an ensemble linearly polarized at $45°$ with an unambiguously defined information. We are going to see later that the simultaneity of uncertainty and defined information forms the basis of the security of quantum cryptography.

The possibility of superposition together with the resulting interference phenomena leads to all the paradoxes and problems of interpretation of quantum mechanics. This, on the other hand, is also the reason for all that is essentially new compared to the classical information science. It is due to these additional capabilities that the use of the term qubit has been adopted into the language for a bit, of which the formation and dynamics can be represented quantum mechanically.

The situation gets even more interesting when the superposition of states of several qubits is considered. A pair of classical bits can have the value combinations "00", "01", "10" and "11". In analogy to the representation above, the basic settings of a pair of qubits can be defined as $|0\rangle_1|0\rangle_2$, $|0\rangle_1|1\rangle_2$, $|1\rangle_1|0\rangle_2$ and $|1\rangle_1|1\rangle_2$. States of this kind are called product states, because the overall state of both qubits can be described as a product of two single states.[2] The two qubits can of course be prepared in any superposition of these basis states. We

2) Compare to Chap. 2.2.

can assume, for instance, that we have a system of two qubits in the state

$$(|01\rangle - |10\rangle)/\sqrt{2}$$

This state means that the two qubits can be found in opposite settings. When we find qubit 1 with the value "0", qubit 2 will have the value "1" and vice versa. When only one single qubit is observed, the value "0" or "1" will show up with the same probability. Under this condition, the qubit has no unambiguous value, and therefore no unambiguous ("pure") quantum state can be assigned. We know a similar behavior from flipping coins. Completely randomly, we see either "heads" or "tails" on top. Still we know that the opposite value is on the lower side. In contrast to this simple classical example, a pair of qubits can be observed under different directions, and thus also in superpositions of the basis states of the single qubits.

Now we take a pair of light quanta, i. e. photons. This pair could be in the state $(|HV\rangle - |VH\rangle)/\sqrt{2}$. When the first photon is measured with horizontal polarization, one knows that the other can be observed with vertical polarization and a measurement will never give the result "horizontal polarization". But when the polarizer is oriented at $45°$ $(= (|H\rangle + |V\rangle)/\sqrt{2})$ and the first photon is detected, the other photon is found at $-45°$ $(= (|H\rangle - |V\rangle)/\sqrt{2})$. As soon as the first photon in the set direction (no matter which direction) is found (and it will be detected with a probability of 50 % along that direction), one knows that the second photon can be measured with an orientation exactly in the opposite direction. This mutual dependency of the two qubits is formally expressed by the fact that these states can no longer be represented as a product of two independent qubit states — therefore, they are not product states.

Schrödinger used the term *entanglement* (in German: Verschränkung) to name this characteristic quantum-mechanical behavior. These nonclassical correlations were "the essence of quantum mechanics" to him, and even 60 years after this statement, more and more new properties of this state are found. The phenomenon known best was shown in a gedanken experiment by Einstein, Podolsky and Rosen, intended by the authors to demonstrate the incompleteness of quantum mechanics. Under the assumption that measurements at entangled particles can be described similarly to classical correlations, they derived contradictions within quantum mechanics. Some time later, however, J. Bell was able to show that an upper limit for classical correlations between measurement results of a system of two particles exists that is broken by quantum-mechanical predictions for measurements of entangled systems (see box "Bell's inequality").[3] We have to depart from the conventional views due to the fact that many experiments indicate that this limit is really broken. After a final confirmation of the experiments, we have to accept either that the result before the measurement of a single particle is by principle undefined, or that this measurement result is defined by "spooky influences", as Einstein put it with a shudder, even over large distances. Maybe both explanations are correct, even though they do not sound very plausible for our classical understanding (Wheeler and Zurek (1983)). No matter in which way this discussion that is so important for the physical understanding may end, the quantum information science uses the features of entangled systems albeit their philosophic consequences.

Bell's inequality

When two entangled photons originally prepared in the state $|\Psi^+\rangle$ are measured, we can observe perfect correlations. A measurement of the polarization of one of the two photons gives with the same probability the result "vertical" or "horizontal" polarization. Immediately after the measurement of one of the two particles (for example with vertical polarization), we know that the other photon can be found under horizontal polarization, even when the two photons are far apart from each other. The same holds for any possible analysis direction.

Einstein, Podolsky and Rosen realized already in 1935 that this attribute of quantum-mechanical systems is incompatible with the condition of locality and reality. Here, locality means that no physical effect can spread from Alice's apparatus faster than the speed of light and possibly influence the result of a measurement from Bob. Reality means that Alice can conclude about the result of Bob from her measurement, because it was in principle already predetermined. These two conditions are always fulfilled in classical physics, but they are not compatible with the basic principles

---

3) See also Sections 2.5 and 7.2.

of quantum mechanics in the way they were introduced by Niels Bohr. For a long time, it remained unclear if quantum mechanics is not yet a complete description, or if we have to give up one of these so familiar conditions (or even both). Only in the year 1964 did John Bell show that there are really measurements that should give an answer to this, when analyzed. When the requirements of locality and reality are also true outside of classical physics, the correlations between certain measurements have to comply with an inequality, while at the same time quantum mechanics predicts exactly for these measurements that this inequality is broken.

For this analysis, Alice and Bob measure independently one of the photons of an entangled pair, each along two possible directions. Alice orients her polarizer along the angles α and β, and Bob chooses between the angles β and γ. Following the original work from Bell, E. Wigner showed that the following inequality must be fulfilled under the assumption of locality and reality:

$$N(1_\alpha, 1_\beta) \leq N(1_\alpha, 1_\gamma) + N(1_\beta, 0_\gamma)$$

Here, for example $N(1_\alpha, 1_\beta)$ is the number of events for which Alice gets from her polarization analysis the result "1" (parallel orientation) under the angle α and Bob the result "1" under the angle β and so on.

According to quantum mechanics, the number of these events is given by $N(1_\alpha, 1_\beta) = 1/2 N_0 \cos^2(\alpha - \beta)$ or $N(1_\alpha, 0_\beta) = 1/2 N_0 \sin^2(\alpha - \beta)$, respectively, where N_0 is the number of pairs emitted by the source. The inequality above, however, is significantly broken when the angles are chosen so that $(\alpha - \beta) = (\beta - \gamma) = 30°$.

This breaking of the inequality was already experimentally observed many times. Even though there is still a last loophole open, some new but quite implausible mechanisms would be necessary in order to justify the assumption of locality and reality at the same time. Starting from the position that the conventional quantum mechanics is correct, the degree of the breaking of Bell's inequality is a direct measure of the entanglement of the observed photon pairs, and therefore it can be used as a security check in quantum cryptography.

In analogy to the four unambiguously distinct (orthogonal) classical product states, four orthogonal entangled states exist, the so-called
Bell-states:

$$|\Psi^+\rangle = 1/\sqrt{2}\,(|0\rangle_1\,|1\rangle_2 + |1\rangle_1\,|0\rangle_2)$$
$$|\Psi^-\rangle = 1/\sqrt{2}\,(|0\rangle_1\,|1\rangle_2 - |1\rangle_1\,|0\rangle_2)$$
$$|\Phi^+\rangle = 1/\sqrt{2}\,(|0\rangle_1\,|0\rangle_2 + |1\rangle_1\,|1\rangle_2)$$
$$|\Phi^-\rangle = 1/\sqrt{2}\,(|0\rangle_1\,|0\rangle_2 - |1\rangle_1\,|1\rangle_2)$$

These four states also form a basis for the description of two-particle
states. In contrast to the four product states, the entangled states can
be transformed into each other by manipulation of only *one* of the
two particles. When, for example, the value of the second qubit of
the state $|\Psi^+\rangle$ is flipped, which means that the value is changed from
0 to 1 or from 1 to 0, the state $|\Phi^+\rangle$ is obtained. A phase change,
that is a change $0 \to 0$ or $1 \to -1$, transfers this state once again to
$|\Phi^-\rangle$, etc. For the product state, in contrast, both qubits have to be
manipulated, in order to get, for example, from $|0\rangle_1|0\rangle_2$ to $|1\rangle_1|1\rangle_2$.

In summary, we can define three essential attributes of entangled
pairs of qubits:

- perfect correlation between the measurement results from the
  particles of a pair, even though the results for the single measurements are completely undetermined,

- different statistics for measurements at entangled or not entangled pairs and

- the possibility to transform the four basis states (Bell states)
  into each other by manipulation of only one of the two particles.

These three attributes are the basis of the important methods of quantum communication with entangled particles.

Quantum cryptography

For a long time people have been trying to transmit messages in a
way that besides the sender and receiver no unauthorized third party
gets to know anything about it. The classical cryptography provides

many tricky methods, which nowadays are not only used for military purposes, but also for economic purposes. One of these methods, the so-called one-time-pad encoding, could in principle be safe against eavesdroppers. Such a high security is achieved by encoding each symbol of a text with its own random key symbol. As shown by C. Shannon, a text coded in this way can be decoded only by using the right key symbols. An eavesdropper cannot understand the message unless he knows the key.

Sender and receiver, let us call them Alice and Bob, can safely exchange a message only when the key is actually secret. Therefore, they are facing a very difficult problem — before the actual transmission of a message, the key has to be transferred between Alice and Bob in an absolutely secure way. In the classical world, a measurement of transferred signals can always be carried out in a way that Alice and Bob cannot realize the occurrence of the bugging. Even a CD brought by a courier could have been read without leaving any traces behind. So how can Alice and Bob be sure that the key is secret?

Now this is the point where quantum cryptography can be employed in the way it was proposed for the first time by Ch. Bennett and G. Brassard in 1984 (Bennett and Brassard (1984)).[4] Quantum objects and in particular superposition states, as was mentioned before, are so sensitive that a single observation can change them completely. When quantum objects are transmitted for the key transfer and key generation, the eavesdropper disturbs this process in such a way that he can easily be uncovered.

For the secure key generation and transmission, Alice sends, for example, polarized photons to Bob. Just by chance she chooses one of the four possible polarization directions H, V, $+45°$ or $-45°$. Bob also switches his polarization analyzer randomly between the H/V basis and the $+45°/-45°$ basis and he tells Alice when and under which basis settings he has detected a photon (but not what result he has obtained). Alice checks in her list the detection times (many photons could have been lost along the way due to absorption) and then she tells Bob at what time they were using the same basis. Since, for example, a horizontally polarized photon sent by Alice in the H/V basis can *only* be detected in the H-output, there is an unambiguous relation between the value that is set by Alice and the value detected

---

4) See also Section 2.6.

by Bob in the case that both were using the same basis. Therefore, they can use this bit sequence as a key.

But how can Alice and Bob be sure now that this process was not tapped? An eavesdropper in a quantum transmission corresponds to a measurement at which an eavesdropper tries to determine the polarization of the photon sent by Alice and to pass on a correspondingly polarized photon to Bob. When the measurement apparatus of the eavesdropper has the same orientation as the one from Alice, he will observe the correct bit and correspondingly send it on to Bob. But when this apparatus is oriented in the other basis, all the information about the original state of the photon is lost. So when the observed polarization is sent to Bob, he could possibly detect a wrong bit on his side. Figure 6.2 shows a typical case in which the polarization of the photon sent by Alice is different from the polarization of the photon detected by Bob, because of the attack of an eavesdropper. In total contrast to the transmission of classical signals, the tapping inevitably causes errors in the key. With the comparison of a few bits in the key, Alice and Bob can therefore find out immediately if the key transmission went safely and undisturbed. The key obtained in this way can ideally be used for the one-time-pad coding and it guarantees for the first time a truly secret communication.

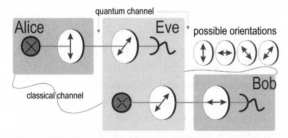

Fig. 6.2 When Eve (for eavesdropper) tries to tap the transmission of qubits between Alice and Bob, she will cause errors in the key bits. When an eavesdropper breaks the line, Bob (as in this example) might detect a horizontally polarized photon, even though Alice has sent a vertically polarized photon.

Our entangled world has yet another possibility ready for a safe key exchange. From a source of entangled photons, one is sent to Alice and the other one to Bob, so they can measure these photons along different directions. When they have oriented both of their analyzers by chance along the same direction, for example along V/H, Bob always observes a horizontally polarized photon when Alice has observed a vertically polarized photon and vice versa. These perfect correlations can be used for the generation of the key. But the times when they were using different settings do not need to be ignored. These results can be analyzed using a Bell inequality. The breaking of the inequality is possible only when the observed photon pairs were really entangled. When an eavesdropper tried to measure the state of one of the two photons, she inevitably would have destroyed the entanglement. Since such a measurement had inevitably transferred the two photons to a product state, Bell's inequality could not be broken and Alice and Bob would have discovered the attack of the tapping person with certainty.

Quantum teleportation

We assume that Alice wants to send an object to Bob. When she is not able to send the object directly to Bob, she can still, within classical physics, determine precisely all properties of the object and pass this information on to Bob. Under the premise of having the right technology, he can then make a perfect copy of the original object. Unfortunately this strategy does not really work. When trying to measure smaller and smaller parts of an object, for example its atoms or molecules, quantum mechanics says that from a certain point on, not all the properties can be determined absolutely precisely. It is because of this that a transmission becomes impossible.

There is still another possible solution. In principle, one only has to make sure that Bob's object at the end of the transmission has exactly the same properties as the one of Alice in the beginning. For this, it is not necessary that the two (or anyone else) know these properties. Ch. Bennett, G. Brassard, C. Crepeau, R. Joszap, A. Peres and W. Wooters showed in 1993 that a solution of this problem is possible by using the entanglement between quantum objects (Bennett et al. (1993)).[5]

We have discussed above that two entangled particles or qubits have no well-defined state for themselves. We know, however, that

5) Compare to Section 7.2 and 9.10.

for the so-called antisymmetric entanglement given above, the two particles will always be polarized in the opposite direction. Before the start of the teleportation, Alice and Bob share an entangled pair of qubits in the state $|\Psi^-\rangle$ between each other (Fig. 6.3). Alice can now carry out a measurement on hers and the qubit to be teleported, which provides no information at all about the quantum state of these two qubits. The so-called Bell-state measurement projects the qubit pair at Alice to entangled states, and thus it is not giving any information about the states of the single qubits, but rather about the correlations between them. We assume that the state $|\Psi^-\rangle$ was found as one out of four possible results. Then we know that the qubit to be teleported is oriented opposite to the state of the second qubit from Alice. Since this state is also opposite to the state of Bob's particle, the original state must be equal to that of Bob's particle. When Alice finds both of her qubits in one of the other four orthogonal, entangled states, Bob has to carry out one of three defined manipulations on his qubit, in order to transform it to the correct initial state.

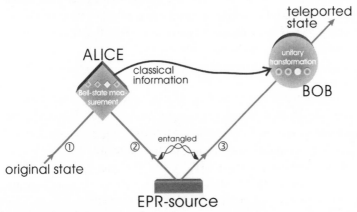

Fig. 6.3 Scheme of quantum teleportation. When Alice and Bob each receive one particle of an entangled pair from an EPR-source (for Einstein–Podolsky–Rosen), the quantum state from particle 1 can be transferred (teleported) to particle 3.

Alice will not find out anything about the state to be teleported with the measurement. However, since a correlation chain between the settings of the three particles is formed, Bob, after receiving the measurement result, can transfer his qubit to the right state. In contrast to the classical case, no copy is made, because the first particle loses its attributes with the Bell-state measurement and it is left behind as a virtually empty, shapeless holder. Also, *no* matter is transported during teleportation. To transfer the state successfully from one particle to another, it is sufficient to transmit the information about their relative attributes.

Quantum computer

A classical computer is a machine that generates, based on a specific input, an output as a result of the arithmetic operations. Input and output for a classical computer consist of bits that always have a well-defined value. The computer operations are expressed by Boole's algebra. The physical realization of such a computer should be thought of as a classical machine, in which the values of the bits are in general given, e. g., by electrical potentials of storage elements.

In a quantum computer[6], we have a number of qubits as inputs and outputs, and the computer generates the output from the input by using quantum-mechanical operations (Fig. 6.4).[7] The operation of such a computer can be formally expressed as

$$|output\rangle = U|input\rangle$$

where the input state in the simplest case could be the direct translation of the classical input, for example $|input\rangle = |0110\ldots010\rangle$, and U, the calculation, is a so-called unitary transformation. From this input state an output state is formed by the operation, and this output state of course corresponds exactly to the bit sequence of the classical result. We have carried out this specific calculation with our quantum system, but obviously nothing has been gained yet.

However, we have not made any use of the characteristic attributes of quantum mechanics, like for example superposition, interference and finally the entanglement between qubits.

The superposition allows us to construct the input state as a superposition of many single classical possibilities:

$$|input\rangle = \{|input\rangle_1 + |input\rangle_2 + \ldots |input\rangle_N\} \Big/ \sqrt{N}$$

6) Compare to Sections 4.5, 7.4 and 9.1.
7) See also Section 4.5 and Chapter 7.

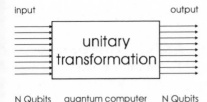

input   output

unitary
transformation

N Qubits   quantum computer   N Qubits

Fig. 6.4 Principle of a quantum computer. The algorithm is represented by a unitary operation that connects the N input qubits with the N output qubits.

With this input state, we get as an output a superposition of the results of all single calculations that were carried out by the quantum computer in parallel:

$$|output\rangle = \{|output\rangle_1 + |output\rangle_2 + \ldots |output\rangle_N\} \Big/ \sqrt{N}$$

With the readout of the output register, always only one of these results with the corresponding — in this case the same — probability is observed. Therefore, the calculation process has to be repeated many times, and still nothing has been gained so far.

The essential point though is that by using the massive quantum parallelism[8], properties can be found that are common to all results, without the need to carry out the calculation for each single input state separately. One possibility, for example, is the determination of the period of a function by the quantum Fourier transformation.[9] When the period of the function is M, all output values at the distance M will have the same value, which means that the same quantum state is obtained for many single calculations. Let us assume that the quantum computer carries out the calculations simultaneously for so many input states that already several periods of the function are covered. Many components of the output superposition will then have the same value and they can interfere after the final Fourier transformation. With a far-ranging destructive interference, the output state is reduced to very few components and the period M can be deduced from the measurement of these. Instead of performing many single calculations and a classical Fourier transformation, just

8) Compare to Section 7.5.
9) See also Section 7.5.

a few quantum parallel-calculations are sufficient for the solution of the problem.

These facts made it possible for P. Shor to formulate the so far most efficient quantum algorithm.[10] The difficulty to factorize a large number grows exponentially with the number. When the number of digits of a number to be multiplied is doubled, the computation time and memory space needed is also doubled. In contrast to this, the computation time for a prime factorization grows exponentially. It is possible, for example, to calculate quite fast how much 107 times 53 is, but it takes significantly longer to determine the prime factors of 5671. One can find without any problems a number so large that it cannot be split into prime factors within a reasonable time. Shor used a known theorem from the number theory for his quantum algorithm, according to which the factorization of a number can be put back to the finding of a period of a function (Shor (1997)).

6.3 Experimental realizations and developments

Great advances in experiments with single quanta have been made over the past decades. Important results regarding the trapping and cooling of atoms and single ions, the resonant amplification of the atom–light interaction, and experiments with single photons have been achieved by means of improved lasers and more efficient vacuum apparatuses. In particular, the experiments with single photons form a solid basis for the development of experimental quantum communication, due to the possibility to send light quanta over large distances. Building a quantum computer is considerably more difficult.[11] The high sensitivity of a quantum state against external influences is of course especially relevant when working with many qubits. In the following, we will give a short overview of the status of the experiments for the realization of the fascinating suggestions from the field of quantum information transfer and processing.

For quantum cryptography, the transfer of single qubits is already sufficient. This is why well-advanced systems for the secret key generation already exist, starting with the first experiments in 1991. The development goes towards apparatuses that are easy to operate, and

10) See also Section 7.5.
11) Compare to Sections 4.6 and 7.6.

that can work independently for a longer time with just as little servicing by the user as possible. A short time ago huge optical tables in air-conditioned labs were necessary for good noise behavior of quantum cryptography, but today systems of the size of a television are available, which can be carried anywhere. The group of N. Gisin at the University of Geneva succeeded to realize quantum cryptography over a distance of 67 km. The apparatuses of Alice and Bob were placed in two post offices and the photons were sent through a glass fibre of the Swisscom (Muller et al. (1997)).

Essential for the functioning of quantum cryptography is the condition that the "quantum line" is not interrupted by amplifiers or switches. Only the completely undisturbed forwarding of photons guarantees a correct generation of the key. But the rates at which key bits are generated drop because of absorption in the glass fibers, until finally the detector noise causes too high error rates. Further developments of detectors and other components suggest transfer distances of about 200 km appear to be realistic. To bridge larger distances, the team of R. Hughes at the National Laboratory in Los Alamos, USA is working on so-called free-space quantum cryptography. (Buttler et al. (1998)). In this case, the photons between Alice and Bob are to be transferred by using satellites that are close to the earth. Under the condition that the eavesdropper cannot have any access to the satellite, there is even the option that Alice and Bob exchange keys over the satellite. From the two single keys a new key that is as secret can be generated. This allows Alice and Bob the secure communication over, in principle, any large distance.

In our labs we are trying to build sender and receiver modules as compact and stable as possible. The complete module should be reduced at least to the size of a modem and it should be possible to connect it directly to a desktop computer, so that the user can really enjoy all the advantages of quantum cryptography. Without any intervention from outside, readjustments (rarely necessary) are carried out automatically. The secure key generation runs in the background for a network of users. In the first prototypes laser diodes are used as photon sources. Their short nanosecond pulses are correspondingly polarized and coupled into a fiberglass line after attenuation. Highly sensitive detectors register single light quanta in Bob's module. In the case of a direct-sight connection between Alice and Bob, the transmission of the photons can also be carried out through telescopes according to the free-space concept, which will be simpler and also cheaper in many cases.

The single pulses contain one photon with a very small probability, but it is even more unlikely that they contain more than one. The security, which means the key rate of the system, can be increased when true single-photon sources are used, or even better, sources of entangled photon pairs.

In the process of parametric fluorescence, when an ultraviolet (UV) laser beam passes through an optically nonlinear crystal, a fraction of the UV photons is transformed into pairs of red photons. The two red photons that are generated in such a process fly into very specific directions, because of the conservation of momentum and energy (Fig. 6.5). Therefore, when a photon is detected in one outgoing beam, one knows that one photon was also emitted into the other beam. We thus have an almost ideal single-photon source. With a suitable alignment, an entanglement of the polarizations of the photons can also be created (Kwiat et al. (1995)).[12]

parametric fluorescence (type II)

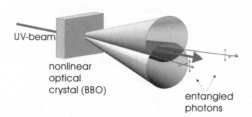 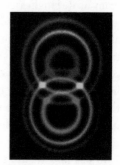

Fig. 6.5 Parametric fluorescence of the type II. An optically nonlinear crystal is irradiated with an ultraviolet laser beam. Two photons with different polarization are emitted with a low probability. Under certain emission directions, these two photons are entangled with each other (lines of intersection of the two cones). The photo on the right shows the emitted photons for three different wavelengths, photons in the crossing points of the central circles are entangled.

In our experiments on quantum cryptography[13] with entangled photons (Jennewein et al. (2000)), one of a pair of photons was sent to Alice and the other to Bob (Fig. 6.6). We were using in each case

12) For the creation of entanglement se also Section 9.7.
13) Compare to Section 2.6.

500 m long glass fibers that were laid in the basement corridors of the University of Innsbruck. The measurement apparatuses of the two users were located in labs at the opposite sides of the university campus and about 400 m, as the crow flies, apart from each other. Alice as well as Bob had fast quantum-mechanical generators for random numbers that predefined the analysis direction. These were switched with a frequency of 10 MHz by means of electro-optical switches. The directions (H/V and $30°/120°$ at Alice's, H/V and $-30°/60°$ at Bob's) were chosen so that an analysis according to Bell's inequality, in the way introduced by E. Wigner (Wigner (1970)), would provide maximum sensitivity towards tapping attempts.

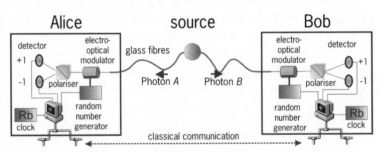

Fig. 6.6 Experimental scheme of quantum cryptography with entangled photon pairs. Polarization-entangled photons are generated in a source using parametric fluorescence and led to Alice and Bob over glass fibers. The two users, who are about 400 m apart from each other, measure independently of each other the polarization of the photons, each along two possible directions. After 1 s of measuring time, the information about the data obtained in this way is exchanged over the classical channel and a secure key is extracted. The security of the key generation against eavesdroppers is checked with the breaking of a Bell-type inequality.

The time, together with the analysis direction and the measurement results, were recorded independently by Alice and Bob in blocks of 1 s. Afterwards, the information about detection time and direction was exchanged over the classical channel, in our case an Ethernet connection between the measurement computers, then the key was generated and the security was checked by an analysis based on Bell's inequality.

In our experiment on quantum teleportation (Bouwmeester et al. (1997)) we used a pulsed UV-laser beam for the generation of photons with parametric fluorescence (Fig. 6.7). In some pulses two photon pairs were generated, of which the entangled pair was shared between Alice and Bob. The photon that was supposed to carry the qubit to be teleported was prepared as a single photon by detection of a fourth photon in the way described above. A specific qubit can be set for this photon through a polarizer. It was then the job of Alice and Bob to transfer the state of this photon; finally it was checked by polarization analysis if and how well the teleportation was working.

The greatest challenge in this experiment was the Bell-state analysis. In the course of this analysis, the two photons from Alice are measured together in such a way that they are projected onto an entangled state. It is not sufficient to measure the photons simultaneously. In principle, an interaction between qubits that are to be analyzed is necessary, in the same way as is needed for the creation of entanglement and later for the quantum-logical operations of the quantum computer.

In the experiment we were using interference effects between the two photons. Though the efficiency is reduced to half with an interferometric analysis, the good quality of this analysis cannot be matched by any other method at the moment. For this purpose, the two photons are superimposed at a beam splitter, a half-silvered mirror. When the incoming beams are well aligned, it is no longer possible to decide which photon hit the beam splitter from what direction and thus interference occurs. When the two photons are in an antisymmetrically entangled state $|\Psi^-\rangle$, they will also behave at the beam splitter in just "the opposite" way. This means that they will leave the beam splitter always through the opposite outputs. In the other three possible cases, both photons will always go through a common output.

For the first demonstration of teleportation, we will restrict our analysis to one single entangled state. When Alice observes a photon in each one of the detectors behind the beam splitter, she registers the antisymmetric, entangled state. Since the entangled pair from Alice and Bob was also prepared in that state, we know that Bob's photon has already the state of the qubit to be teleported, and therefore Bob can give this photon free for the polarization analysis to demonstrate the quality of the teleportation.

The absorption of photons along the fiberglass line sets an upper limit for the distance between Alice and Bob, similar to the situation in quantum cryptography. An entanglement between two photons

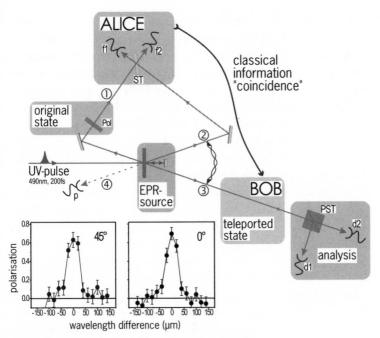

Fig. 6.7 Experimental setup of the quantum teleportation experiment. A short UV pulse irradiates a nonlinear crystal and generates four photons in different directions. When photon 1 and photon 2 overlap at the beam splitter (ST), which means that the path difference equals 0, the quantum state (here the polarization) of photon 1 can be teleported to photon 3. The quality of the transmission is checked by means of a polarization analysis at the polarizing beam splitter (PST). The measurement series show that the polarization could be transferred with an accuracy of about 68 %.

has been already proved for distances of 10 kilometers at the University of Geneva.

While our first experiment was teleporting a qubit from one photon to another, recently a continuous quantum state was successfully transferred. In an experiment at Caltech, USA, the properties of a light field could be transferred to another one (Furusawa et al. (1998)). After the first generation of an entanglement between atoms, the transfer of quantum attributes from one atom to another will quite soon be possible as a next step. This will certainly not be sufficient for the realization of science-fiction dreams, but such an experiment forms the basis for a quantum memory and a quantum computer.

The parametric fluorescence is at present essential for the generation of entangled photon pairs in the way they are needed for our experiments. The generation of entangled particle pairs is still a great challenge in experimental quantum optics. Spontaneous transformation processes like the parametric fluorescence or the dissociation of molecules have been used for quite some time. However, the low efficiency and the fact that the emission time is in principle uncontrolled is a problem for some applications. Quite recently, researchers at the ENS Paris and at the NIST in Boulder, USA, succeeded for the first time to create an entanglement of the states of atoms and ions (Hagley et al. (1997), Turchette et al. (1998)). Though the quality of the entanglement between atoms and ions is not as good as for the photon pairs, the first step on the way towards quantum-logical operations for future quantum computers has been made.

For the implementation of the quantum computer, we assume that the qubits are available in a suitable form as a series of atoms, ions or even nuclear spins. It is possible then to break down the whole unitary transformation U into single steps. These single operations are carried out by so-called quantum gates, which couple two qubits at a time with each other. The coupling has to carry out conditioned logics — operations in which the change of one qubit depends on the state of the other qubit (similar to a logic XOR connection of the conventional computer). Two qubits, for example, can be entangled with each other by means of this operation plus the manipulation of single qubits, and in reversal to this operation they can also be transferred into a product state again. Any possible unitary transformation and calculation operation can be carried out when many two-qubit and one-qubit operations are strung together.

How can such a quantum gate be realized? There are several suggestions in the literature that describe strong couplings, for example between quantum dots, between the nuclear spins of the atoms of a molecule, or between the ions inside a trap. First mini-quantum algorithms have been already carried out with two and three qubits using the nuclear spin resonance of molecules (Cory et al. (1997), Chuang et al. (1998)). Important within this context was in particular the high technological standard of commercial nuclear spin resonance apparatuses. The poor scalability of these systems had disadvantageous effects. Since all qubits sit on one molecule and a coupling between qubits on different molecules is not possible, the step towards increasing the number of qubits is very difficult. This enlargement appears to be easier for a series of nuclear spins in the solid state, which are coupled over electronic states (Kane (1998)).

A very promising idea for the implementation of a quantum gate uses chains of ions that are stored in a trap (Cirac and Zoller (1995) and Fig. 6.8). The change of a qubit can cause a change of their oscillation behavior. For example, all ions of a chain will start oscillating together when the state of one of the ions is inverted with a certain laser pulse that is directed at it. When, on the other hand, another ion is irradiated with a second laser pulse, its state will change depending on the situation, if the chain oscillates or is at rest. Several labs worldwide are now trying to cool such a chain to low temperature so that really all the ions are at rest. Recently, the teams of NIST in Boulder and University of Innsbruck managed to implement the first two-bit operations and could run fundamental quantum algorithms like teleportation and factoring.

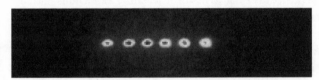

Fig. 6.8 Fluorescence from an ion chain. Ions are stored in a linear ion trap and irradiated with laser light for fluorescence. When these ions are slowed down to an absolute stop, they can be used as the qubits of a quantum computer.

It is not just the handling and control over many qubits that causes problems. In the ideal case, the qubits of a quantum computer only couple among each other, but not with external systems or particles. When an interaction or coupling with the outside world takes place, the state of a qubit is disturbed and in the worst case completely deleted. The loss of quantum information to uncontrollable particles is named decoherence.[14] The average time that passes until the quantum information is destroyed by scattering, collisions or other processes is the decoherence time. This time is shorter the stronger other systems interact with the qubit. When many other particles

14) See Chapter 8.

couple to a qubit, for example in the solid state, this time (less than a nanosecond) is much shorter than for the case of a single ion in the high vacuum of an ion trap chamber (more than a second). But when it is about an entangled quantum state consisting of many qubits, the disturbance of just one of these qubits is enough to delete the information of the overall state. The decoherence time is exponentially lowered — longer calculations with many qubits are therefore impossible. Fortunately, Shore and Steane have discovered methods for quantum error correction. In this case, the quantum information of a qubit is distributed over several qubits, so that the disturbance of one single qubit can be reversed. It is because of this that there are still good chances to solve complex calculations with a quantum computer one day.

6.4 Summary

The fundamental effects of quantum mechanics like superposition and entanglement form the basis for fascinating extensions and improvements of conventional methods for information transfer and processing. Quantum cryptography makes an absolutely safe communication possible for the first time. The first commercial devices have proven their efficiency already outside of protected laboratory conditions. In extended fiberglass lines, secret key information has been generated over a distance of 67 km. The next generation of these devices should regularly guarantee safe data transmission in computer networks.

Quantum teleportation makes the transfer of quantum attributes possible, and with this, in principle, the reconstruction of an object at a far distant place — also within quantum physics. Even though this will certainly not be used for journeys to foreign planets, this idea forms the basis for important elements of quantum information science. The storage of quantum attributes, efficient communication over noisy lines and data transfer between quantum computers are based on quantum teleportation.

The quantum computer is probably the most fascinating innovation. Even though the basic idea is actually similar to the structures of the classical computer, the superposition principle and quantum interference effects make radical extensions of conventional algorithms possible. It is not yet foreseeable, which problems could better be solved with quantum algorithms. The first examples, like the prime factorization show that calculations that take centuries on traditional computers, could be finished on a quantum computer within a few minutes.

Still, this quantum computer exists only on paper. Different methods like nuclear spin resonance or the spectroscopy at ion chains and quantum dots are being investigated at present and they are being decisively developed further towards possible applications. It is not yet clear which one of the systems suits best and will eventually make the efficiency of quantum information science usable. The solution is not just around the next corner, but the way to get there will also bring valuable technological developments for other branches of science and industry and perhaps even some better understanding of the so often paradoxical and counterintuitive quantum phenomena.

References

– C. H. Bennett, G. Brassard (1984), *Proc. IEEE Int. Conf. Computer Systems and Signal Processing*, Bangalore, India. IEEE, New York, p. 175.
– C. H. Bennet, G. Brassard, C. Crépeau, R. Josza, A. Peres, W. K. Wootters (1993), *Phys. Rev. Lett.* **70**, 1895–1899.
– C. H. Bennett (1995), Quantum Information. *Physics Today*, October, p. 24.
– D. Bouwmeester, J.-W. Pan, K. Mattle, M. Eibl, H. Weinfurter, A. Zeilinger (1997), *Nature* **390**, 575–579.
– D. Bouwmeester, A. Ekert, A. Zeilinger (Hrsg.) (2000), *The Physics of Quantum Information*, Springer-Verlag, Berlin.
– W. T. Buttler, R. J. Hughes, P. G. Kwiat, S. K. Lamoreaux, G. G. Luther, G. L. Morgan, J. E. Nordholt, C. G. Peterson, C. M. Simmons (1998), *Phys. Rev. Lett.* **81**, 3283–3286.
– I. L. Chuang, N. Gershenfeld, M. G. Kubinec (1998), *Phys. Rev. Lett.* **80**, 3408–3411.
– J. I. Cirac, P. Zoller (1995), *Phys. Rev. Lett.* **74**, 4091–4094.
– D. Cory, A. Fahmy, T. Havel (1997), *Proc. Nat. Acad. Sci.* **94**, 1634.
– A. Furusawa, S. Sørenson, S. L. Braunstein, C. A. Fuchs, H. J. Kimble, E. S. Polzik (1998), *Science* **282,** 706.

- E. Hagley, X. Maître, G. Nogues, C. Wunderlich, M. Brune, J. M. Raimond, S. Haroche (1997), *Phys. Rev. Lett.* **79**, 1–5.
- T. Jennewein, Ch. Simon, G. Weihs, H. Weinfurter, A. Zeilinger (2000), "Quantum Cryptography with polarization entangled photon pairs", *Phys. Rev. Lett.* **84**, 4729–4732.
- B. E. Kane (1998), *Nature* **393**, 133–138.
- P. G. Kwiat, K. Mattle, H. Weinfurter, A. Zeilinger, A. V. Sergienko, Y. H. Shih (1995), *Phys. Rev. Lett.* **75**, 4337–4341.
- H.-K. Lo, S. Popescu, T. P. Spiller (eds.) (1998), *Introduction to Quantum Computation*, Clarendon Press, Oxford.
- A. Muller, T. Herzog, B. Huttner, W. Tittel, H. Zbinden, N. Gisin (1997), *Appl. Phys. Lett.* **70**, 79.
- P. W. Shor (1994), *SIAM J. Comput.* 26, 1484. Another important quantum algorithm comes from L. Grover (1997), *Phys. Rev. Lett.* **79**, 325–328.
- Q. A. Turchette, C. S. Wood, B. E. King, C. J. Myatt, D. Leibfried, W. M. Itano, C. Monroe, D. J. Wineland (1998), *Phys. Rev. Lett.* **81**, 3631–3634.
- J. A. Wheeler, W. H. Zurek (eds.) (1983), *Quantum Theory and Measurement*, Princeton University Press, Princeton. A comprehensive collection of the original articles on entanglement, Bell's theorem and on the interpretations of quantum mechanics.
- E. P. Wigner (1970), *Am. J. Phys.* **38**, 1005–1010.

7
Quantum computers — the new generation of supercomputers?

Reinhard F. Werner

7.1 Introduction

When Peter Shor published his quantum algorithm for the factorization of large numbers in 1985 and the sensational news reached the press, readers might have asked themselfes if this was a serious publication. Calculations made faster with quantum particles? Was "quantum" not actually a popular prefix for products from American quacks? Or for the better informed: Was quantum mechanics not actually this paradoxical branch of modern physics, that even Einstein and Schrödinger never really got to like? And just these paradoxes are supposed to drive a computer!

In this chapter I would like to follow this kind of question and show which principles of quantum theory make a quantum computer possible, but also in which sense one could say about a computer, which has not been built yet, that it is going to be faster than any present (and future!) machine based on conventional principles. As an example of a quantum algorithm, I will present the Shor algorithm, which still promises the most dramatic speed-up of all known quantum algorithms. Peter Shor's achievement was only possible by combining principles and results from physics, information science and pure mathematics (number theory). I hope that some of this fascinating combination will become transparent, even though a complete discussion would go beyond the scope of this book.

Entangled World, Jürgen Audretsch
Copyright © 2005 WILEY-VCH Verlag GmbH & Co. KGaA, Weinheim
ISBN: 3-527-40470-8

7.2 Quantum information

In the same way a classical computer can be understood as a gadget for the processing of (classical) information, a quantum computer is processing "quantum information". In this section we are first going to clarify what is meant by quantum information and how this kind of information differs from its classical counterpart.

Quantum mechanics and statistical interpretation

Instead of giving a detailed presentation of the quantum-theoretical formalism, we are going to restrict ourselves to some essential features that can be understood also without the formal machinery. Our starting point is the remark that statements in quantum mechanics are always of a statistical nature. Such statistical statements are well known from classical physics. Therefore, we are first going to recapitulate how a statistical statement comes about.

Statistics and probability theory are typically applied when different results always occur at the repetition of one and the same experiment, but with a *reproducible frequency*. The dice and other games of chance are such a situation. Here, the random events in the sense of probability theory are clearly not just unpredictable, chaotic and muddled, but they are well predictable at least in one respect. This kind of reproducibility can actually be expected only from well-defined situations. The "probability that a stone is heavier than 42 g" is surely not defined well enough: in this case everything would depend on the place where we pick up the stones.

The setting of a defined situation is named *preparation*. Very often this is only considered in a way that certain physical objects are "established", for example, the die in free fall, roulette tables that are turned on, electrons in an accelerator and so on. It is important that a detailed description about the way in which the situation should be set (the "preparation procedure") is always given, because when we want to repeat "the same" experiment, we must have a definition of what actually counts as repetition. This definition of the procedure is a more reliable guideline for the repetition than the demand that all experiments have to be carried out with the same system: the die can get worn out, and the microsystems used in quantum-mechanical experiments, as a rule, can no longer be found afterwards.

Now, a *measurement* is carried out on objects prepared in this way (Fig. 7.1). A certain procedure is also described for this measurement, so that we are able to carry out a whole series of experiments with always the same experimental setup (preparation and measurement). The relative frequencies of the experimental results are noted down for such a series of measurements, and these are the numbers for which a statistical theory like quantum mechanics can make predictions.

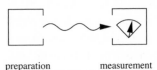

preparation   measurement

Fig. 7.1 Scheme for a statistical experiment of preparation and measurement.

These predictions will of course never be exactly fulfilled. Especially in the case that everything goes according to the rules of probability theory, fluctuations are to be expected, which approach zero for N single experiments like $1/\sqrt{N}$. When this is not the case, or the measurement series shows other rough differences from the typical behavior of random sequences, the measurement has failed. This could be due to any relevant influences from the surroundings (fluctuations in the power supply system, crosswind) that were not properly controlled, and hence the preparation and measurement procedures were not specified well enough. When it is not possible to achieve reproducible frequencies with improved experimental methods, probably the theory is simply not applicable. No final conclusion can be drawn based on a finite series of measurements. There is always some room for interpretations, which among others has led to a still ongoing controversy about the basics of probability theory.[1]

In quantum mechanics we take the possibility of statistical experiments for granted. Moreover, we are going to deal exclusively with

1) The basic positions are the frequency interpretation from R. von Mises (e. g. in *Wahrscheinlichkeit, Statistik und Wahrheit*, Springer, 1936) and the theory of the "subjective probability" from B. de Finetti (e. g. in *Probability Theory*, Oldenbourg, Munich, 1985) in which the probability is defined by the willingness of individuals to make bets on different rates.

the statistical statements (probability predictions) of the theory. This is called the *statistical interpretation* of quantum theory,[2] implying the existence of other interpretations. The debate about the basics of quantum mechanics is indeed even tougher than the debate about the basics of probability theory. Starting from the problematic idea that single quantum particles could be described by a wave function, over theories with hidden variables in the style of Bohm and Nelson (Bohm (1952), Nelson (1966)), there is a colorful zoo of sometimes vehemently defended "interpretations". Fortunately, all these theories agree when it comes to the calculation of probabilities for a concrete experiment. Regarding the question of how often a photon hits the counter in a given experimental setup, it is just irrelevant, if this happens because it was interacting with shadow photons from a different universe (Deutsch), or a quantum potential had led it onto crooked paths (Bohm), or that we just refrain from any "explanation". In this sense the statistical interpretation is the common core of all interpretations. One of the theses of this chapter is that this minimum interpretation is already sufficient to explain quantum information, the quantum computer and all these. A proof of this is already the observation that the researchers in this field are not arguing about the question whether for example a quantum computer could work, in spite of very different positions about the basics of quantum mechanics.[3]

Of course almost nothing is said with the statement that quantum mechanics is a statistical theory, because many classical theories are also of this kind. Exactly within this framework, the difference between classical (also statistical) theories and quantum mechanics can be worked out best. This is going to be the topic of the following paragraphs.

2) Compare to Section 1.6.
3) This is not about objections against the practical feasibility of the construction of a computer from components of a finite quality, but objections from the basics of quantum mechanics. An exception from the consensus maintained here is, e. g., Gerhard t'Hooft, who holds the opinion that physics on the very smallest scales (e. g. the Planck scale of 10^{-34} cm) can get classical again. He thought that he could formulate a criticism of the research program of quantum information theory based on that. Nobody is going to question t'Hooft's competence regarding the Planck scale. Unfortunately, however, he has not given any explanation why these effects on the one hand should result in the massively observed nonclassical correlations over large distances, but on the other hand they should respect a strict causality of the signal.

Quantum information

After having said a few words about quantum mechanics, we now get to the second part of the expression "quantum information", information theory. Since Shannon (Shannon (1948)), this has been an established field of the engineering sciences. A standard task here is the optimized transmission of messages from a sender to a receiver. When picturing this situation we are reminded of Fig. 7.2. Exactly this analogy to quantum mechanics turns out to be fruitful: the sender corresponds to the preparation of quantum systems and the receiver to the measurement. Within this picture, the particles moving from the preparation part of an experiment to the measurement transmit some kind of information. Therefore, as a preliminary definition, we could say: *Quantum information is the type of information that is transferred by quantum particles in a quantum-mechanical experiment from the preparation to the measurement apparatus.*[4]

Fig. 7.2 A teleportation line: translation of quantum information into classical information and back.

Isn't that a little odd? Why couldn't we also define: "Radio-waves information is the type of information that is transferred by radio-waves"? Wouldn't it be necessary then to create for each physical carrier system its own term for information? The classical information theory, at least, does not do this. The simple reason why this distinction is not made and better should not be made anyway is that the information can easily be *translated* between all these physical carrier systems. All the steps on the way from the formulation of this sentence in my mind, to the keyboard, to my hard drive, into the e-mail, over the internet to the publishing house, finally printed as pixels on the page, and in books in order to reach the kind reader are pure routine. Wherever losses occur during such translation steps, they are well understood and they can be controlled again using information theory. In other words, the classical information term is

4) This statement is not all that new. For example at Ludwig, a slogan of axiomatic quantum theory is (freely formulated): "Microsystems are effect carriers". See, e. g., G. Ludwig, "An axiomatic basis for quantum mechanics I", Sect. III, § 4, Springer 1985.

an abstract concept that becomes real with the possibility to translate between all classical carrier systems.

So the preliminary definition of quantum information from above only makes sense when we provide the evidence that this type of information *cannot* be translated into classical information without losses. Fig. 7.2 shows what such a translation should look like. In all the diagrams of this kind, the quantum particles are represented as wave lines and the classical information as straight arrows. The first black box in Fig. 7.2 is therefore a system that extracts classical information from quantum information, or more directly: the first step on a chosen translation line is a measurement. The classical data thus extracted are now transmitted to the second black box, which prepares a new quantum particle. In the course of this process, the classical information can be used in any way. In the case of a functioning translation line, the output would now be a perfect image of the input. For this, it is irrelevant if an "equal" or even the "same" particle comes out at the end. Since we want to stay completely within the framework of the statistical interpretation, we can just demand that with any preparation of the inputs and any measurements at the outputs, all frequencies found with and without translation line are the same. In this case, we also call this *classical teleportation*.[5]

Our claim, therefore, is that this teleportation process is impossible. Of course we could prove this directly as a theorem from the structure of the quantum theory. However, this is not just the easiest way even for physicists who have the necessary knowledge to understand such a line of arguments. In the following paragraph, we will instead make a connection from teleportation to some other problems of which the impossibility is better known (and also easier proven).

At this point we can already make a few general remarks about the term quantum information. At first it should be clear that we are not going to start here with a new kind of quantum theory. When we take our preliminary definition seriously, just about anything to be

---

5) Being reminiscent of science fiction was absolutely intended by the creators of this terminology (see for example
http://www.research.ibm.com/quantuminfo/teleportation).
The attribute "classical" is supposed to distinguish the problem considered here from quantum teleportation or "entanglement-enhanced teleportation", which is a really possible, very fascinating transfer process for quantum information that, in fact, has been already realized experimentally. More details can be found for example in Chapter 6 of this book or at the given web address.

said about quantum information can in principle be found in the text-books of quantum mechanics. It is therefore only a new view, a new way of asking questions of a well-known theory. In particular, one can take traditional questions of the classical information theory and think what kind of new aspects show up in the quantum-mechanical context. This is mainly about questions regarding the transmission. Traditionally, quantum-mechanical formulations of the same questions would be, for example, the question about the transmission of "intact quantum states", about the "coherent" transmission or about the transmission "under conservation of all capabilities to interfere".

All this, as well as the term classical information, has only little to do with the colloquial term "information": this is not about the kind of information searched for at an information desk, but about optimized transfer and storage. In other words, the information theorist makes no contribution to the question whether any "disinformation" is practiced from television channels; his concern is the technical quality of the pictures. With respect to the quantum information, we have to disappoint any expectation that there might be messages with some kind of a quantum content. "Read quantum messages" would probably be anyway just classical messages again.

Another remark refers to the abstraction within the quantum information theory. In the same way in which to and fro conversions between classical carriers can be made, conversions are possible between quantum-mechanical carriers — at least theoretically, and more and more so in the experiment. Typical quantum degrees of freedom under investigation are the polarization of photons, excitations of a radiation field in a cavity, excitation states of atoms and a few more. Already in view of the theoretical possibilities of the conversion, it makes some sense to construct an *abstract quantum information theory*. For such a theory, the question about the physical realization of the systems has as little relevance as for the classical theory, as long as they are quantum systems. Of course, the experimental realizations of quantum theory do not by far match the theoretical considerations yet. Therefore, as a good research strategy one should work at first on problems that can be answered for all suggested realizations at the same time. This would be the purpose of the "abstract" theory. In the same way the smallest possible system of the classical information theory is named a bit, the quantum information theory also has a smallest system, the *qubit* (short for quantum bit).

Copying, teleporting and Bell-type inequalities

In this section we are going to show that the impossibility of the classical teleportation is closely related to the impossibility to copy quantum information (no-cloning theorem (Wootters and Zurek (1982))).[6] At first we have to clarify again what we mean by a copier. This is schematically shown in Fig. 7.3, where we can see an apparatus that takes in a quantum particle and gives out two particles of the same kind. Each copy should statistically be as "good" as the input. When we ignore one of the copies and carry out any kind of experiment with the remaining exit, we will always get the same probabilities as we do when working without the copier. A classical copier fulfils this condition quite well, and in most cases the original can be seen as one of the copies. We now claim that such an apparatus for quantum particles cannot exist.

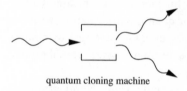

quantum cloning machine

Fig. 7.3 Scheme of a hypothetic copier.

This has obviously massive consequences for the operation of quantum computers. For example, it is impossible to carry out any data backup. Also, a quantum database is quite pointless, because it is impossible to read data without modifying the databank itself. On the other hand, one can also make good use of the impossibility of a copier: a tapping attack on quantum data cannot be made in an unrecognizable way, because otherwise both the eavesdropper and the legitimate receiver would have a copy of the data. This is used for the so-called quantum cryptography, one of the first applications of quantum information theory that was technically realized.

How can we prove that such a quantum copier cannot be built? This is after all a very strong statement, and if the only argument we can give is that such an apparatus is not allowed within the framework of quantum mechanics, one might answer: "Then perhaps quantum mechanics has to be modified". Fortunately, this problem can be di-

6) Compare to Section 2.7.

rectly investigated in experiments without presupposing anything about quantum theory. The relevant experiments are those in which the breaking of Bell's inequality can be proven.[7] In such experiments one investigates correlations between two particles with two different measurement apparatuses being alternately used on each particle. A copier, however, could spare us this choice: according to Fig. 7.4, we could simply copy both particles and then carry out the necessary measurements on the two copies. The actually measured correlations in a real experiment without a copier could also be obtained in this extended experiment due to the copying condition just by ignoring one of the two results on each side. The measured correlations must then be reductions of a common distribution. This is exactly the precondition from which, as is known, Bell's inequalities can be derived. So when a correlation experiment results in a breaking of the inequalities — and this was proven for different systems meanwhile — we can directly conclude that the measurement apparatuses used in that experiment cannot be replaced by one single apparatus, and that a copier consequently cannot exist.

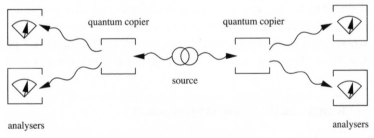

Fig. 7.4 Copier in an experiment on the breaking of Bell's inequalities.

What does this have to do with the problem of quantum information being translatable into classical information, and thus with the possibility of teleportation? Let us assume that we have a teleportation line. We could make good use of the fact that classical information can be copied. So we carry out the measurement of the teleportation line and send one copy of the measurement results to each one of the two preparation apparatuses for the reconstruction of the type we know from the second part of the teleportation line (Fig. 7.5).

7) See also Sections 2.5 and 6.2

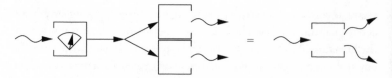

Fig. 7.5 The use of a teleportation device for making copies.

The results of these preparation apparatuses are then quantum copies. Copies and input are statistically equivalent, when the teleportation line operates without errors.

The logical dependence of the three elements discussed in this paragraph is therefore

The impossibility of a copier follows from the experimental evidence for the breaking of Bell's inequality, and from this the impossibility of teleportation. Quantum information, therefore, cannot be translated into classical information. A remarkable point about this argument is that it makes no use of quantum mechanics at all. The no-cloning theorem and the ban on teleportation are thus not simply "casual" attributes of the quantum-mechanical formalism, but they must be true for any theory that gives a local[8] description of Bell-experiments.

7.3 Superposition and entanglement

It is not sufficient to say which tasks are *im*possible with quantum information in order to explain how quantum computers can get to their astonishing efficiencies. The quantum theory should also have something positive to offer. For this, we want to call an attribute of quantum particles to mind, which has been held as the characteristic difference from classical particles since the early days of quantum theory: interference patterns can be generated with quantum-mechanical particles. The classical gedanken experiment for this is

---

8) A locality premise is included in each derivation of Bell's inequalities, which means that the settings chosen on one side cannot be detected as a modification of the probabilities measured on the other side. This is implicitly the case in the argument given here by the precondition that the copiers can only see the particles presented to them and that they work independently of each other.

the scattering at the double slit shown in Fig. 7.6 once again as a reminder. Particles are sent towards a wall having only two holes with a small separation, through which the particles can be transmitted. An interference pattern, represented here as the frequency distribution of the collected particles, results on a collector screen placed behind the wall. The riddle now is how to answer the question: which one of the particles that arrived at the screen went through which hole? On first thought, there would be a simple way to find out something about this: we just hold one hole shut and look at the distribution thus formed. This case is also shown in Fig. 7.6. The interference structure has vanished. The two distributions obtained with only one open hole added together can in no way reproduce the distribution that is measured with two open holes. This is paradoxical, because the particles must have gone through one of the two holes — or perhaps not?

Fig. 7.6 The double-slit experiment.

Quantum mechanics describes these interference patterns without giving any answer to the question of how many particles actually go through which hole. Quantum mechanics also does not deal with the question of what happened with the particles on the way from the preparation to the registration. However, it does explain what happens with the trial to check the path, when any kind of device (e. g. a photoelectric barrier) is mounted at the holes in order to register the particles passing through. The problem in this case is that any such device "disturbs" the particles, which is shown in a modification of the distribution collected on the screen. When the photoelectric barrier is too bright, we can find out through which hole each particle comes, but the interference disappears. In the opposite case of a weak photoelectric barrier that leaves the interference largely undisturbed, only a small amount of "which-path information" is obtained. This difficulty is typical for quantum mechanics. One can hardly tell anything about unobserved quantum particles. Whatever might have happened to the particles on the way from the preparation to the measurement is not a subject of quantum mechanics, and any attempt to invent a plausible story about it leads to strange long-distance effects. This is the crux with many interpretation problems of quantum mechanics.

However, we will not start a debate on fundamental principles, but rather interpret the interferences of quantum mechanics as a chance for new combinations. We cannot keep with the simple alternative "particle went through hole 1 or through hole 2", but will consider additional situations, which are still insufficiently described (because by far too classically) with "particle went through both holes".

The superposition principle as a source of new possibilities

We assume the easiest case of a single bit — a system that has only two possible settings, which we name YES and NO for simplicity reasons.

The possible preparations of such a system (regarding any following statistical experiment) are simply given by the frequency of each of these cases, and thus by the probability of the YES events. This is illustrated in Fig. 7.7. The interval between the preparations "always YES" and "always NO", and thus the amount of all possible preparations is named the space of states of the system. The marginal points have the attribute that they cannot be obtained as combinations of other preparations. Preparations with this attribute are generally named "pure states".

We now take a look at the analogous case of a quantum bit (qubit). We choose the polarization degree of freedom of a photon so that we have something concrete in mind. We can select photons with a given polarization through a filter. In this case, the properties HORIZONTAL/VERTICAL that exclude each other correspond to the classical case of possible settings YES/NO. When a horizontally set polarization filter is combined with a vertically set filter, no photon can get through. On the other hand, two filters of the same type behind each other are (almost) as transparent as one single filter, and this can be taken as a test for the reliability of the preparation. These pure states can be combined statistically as in the classical state, and then we get for example "nonpolarized" photon sources as an equally weighted mixture. This is obviously not the total sum of possibilities. HORIZONTAL

classical bit   quantum bit

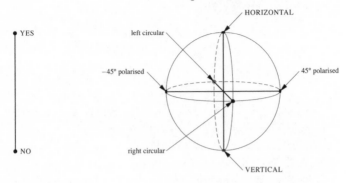

Fig. 7.7 Spaces of states.

and VERTICAL are not the only possibilities to orient a filter: any other angle works as well and obviously each pair of directions perpendicular to each other is as suitable as the pair HORIZONTAL/VERTICAL. Finally there is also the possibility of circular polarizations, and when everything is added together, the full sphere shown in Fig. 7.7 to the right results as the space of states. The exact connection with the superposition is discussed in the box "Superposition in complex numbers".

Superposition in complex numbers

The "pure setting possibilities" of a quantum system form a vector space over the complex numbers. When HORIZONTAL and VERTICAL are such possibilities — following the usual notation, we write $|H\rangle$ and $|V\rangle$, — so are

$$\alpha|H\rangle + \beta|V\rangle$$

equally valid possibilities, where α and β can be any complex numbers with the absolute $= 1$. For waves, this corresponds to the superposition in amplitude and phase. The question of how the probabilities are to be calculated for such a superposition is essential. We will describe this for the simplest case, the superposition of only two possibilities. The possible preparations are then described by a sphere (Fig. 7.7) in the three-dimensional space, where the point for the given superposition has the coordinates

$$x = \text{Re}(\bar{\alpha}\beta)$$
$$y = \text{Im}(\bar{\alpha}\beta)$$
$$z = |\alpha|^2 - |\beta|^2$$

Then obviously $x^2 + y^2 + z^2 = 1$ is true. A measurement, for example, gives the result "yes" at the north pole or "no" at the south pole of the sphere. Any other direction would also result in a possible yes/no measurement.

More generally, the probabilities that occur in an experiment for a superposition $\sum_i a_i |i\rangle$ depend linearly on the products $\bar{\alpha}_j \alpha_i$.

Entangled space

When we apply the combination possibilities extended by super-position to a compound system, we get completely new possibilities of correlations between subsystems.[9] Classically everything is once again very simple: for two bits there are four possible settings, (YES, YES), (YES, NO), (NO, YES), (NO, NO), which correspond to the pure states. These pair-preparations can simply be made in the following way: while a preparation procedure is used for the first particle, an independently running procedure is carried out for the second particle. Each combination is possible in this case, and in addition we obtain again the statistical mixture. The preparation of this mixture should be imagined in the following way: at first a random generator is started and the results are passed on to two preparation apparatuses, which prepare their subsystem depending on this input. Since both preparations are linked by a common history, the result can only be a correlated state.

The random generator could, for example, be realized as a toss of a coin, where the first preparation takes over the result and the second takes the opposite. An evenly weighted mixture of the possibilities (YES, NO), (NO, YES) is obtained in this way. When only one particle is considered, this looks like a simple toss of a coin. But the correlation

9) See also Section 2.2.

becomes obvious from the property that after taking a look at the first particle, the measurement at the second particle can be spared, because the result is already defined.

Quantum states can of course be combined in exactly the same way. States obtained like this are also called *classically correlated* or *separable*. These are the boring preparations. It gets more interesting when the possibilities of *superposition* are used instead. The states that are obtained then are called entangled. These states show now peculiarities similar to the particles at the double slit: instead of simply being able to mix the possibilities (YES, NO) and (NO, YES) as for the classical example, (HORIZONTAL, VERTICAL) and (VERTICAL, HORIZONTAL) can also be brought to interfere. It turns out that the anticorrelation attribute holds not only for the directions HORIZONTAL and VERTICAL, but also for all other orthogonal pairs of directions.

Einstein, Podolsky and Rosen constructed a similar state in 1935 and concluded from the perfect correlations that the polarization attributes should already be well defined at the preparation. The question whether the idea of such attributes, which are also named "hidden variables", can be brought into line with quantum mechanics and nature itself, led John Bell to the formulation of his inequalities — the sharpest conceivable for being the quantitative version of the "paradox" of Einstein, Podolsky and Rosen.[10] Apart from the fact that entangled states are used all the time in quantum-mechanical calculations, the attribute of entanglement was hardly ever a topic until a few years ago. Most of the time they were brought into the debate about the basics of quantum mechanics in order to "... subtly humiliate the opponents of quantum mechanics".[11] There has been a lasting change due to the quantum information theory: entanglement is now becoming a resource, of which the generation is rather a painstaking business. The entanglement can be used to cope with different problems of quantum information processing and it is used up in the course of these processes.

The fact that entanglement really leads to new possibilities can perhaps be seen best in the following table.

---

10) See also Sections 2.4 and 2.5.
11) Freely quoted from a lecture of C. H. Bennett.

system	dimensions of the algebra of observables
1 bit	2
1 qubit	**4**
10 bit	1 024
10 qubit	**1 048 756**

The dimension of the algebra of observables is the number of parameters that has to be given in order to completely characterize a yes/no measurement at a system. The dramatically fast growth of this quantity in the quantum case has long been known in quantum mechanics. The numerical simulation of systems of quantum-mechanical spins, for example, gets to its limits much faster than the corresponding problem for classical spins. This growth also plays an important role in quantum information theory. Assumptions obtained from numerical studies have turned out to be misleading already several times. The numerical calculation can usually only handle systems with very few qubits, not enough to be able to develop reliable intuitions. The development of new mathematical techniques is required in order to overcome these difficulties. This, however, should be worth the effort, because it is exactly the growth of dimensionality that makes quantum mechanics so difficult, but it will actually be used in quantum computers. The first one to realize this possibility was Richard Feynman (Feynman (1982)): When quantum systems are so notoriously difficult to simulate, he said, why shouldn't it be possible to use exactly a quantum system to solve this problem?

7.4 Quantum computer and complexity

What a quantum computer actually is should be clear to a great extent with the comments made so far.[12] The *registers* of such a computer are not memory places for classical yes/no bits, but quantum systems or degrees of freedom of a quantum system. The state of such a register made of several bits will, in general, be entangled to a high degree.

The *gates* of a classical computer are the elementary switching operations, for example the AND operation, which calculates a new bit

---

12) See also Section 4.5, 6.2 and 9.1.

from two bits. It is typical for gate operations that they concern a few register bits, while all calculations can be made with these operations. This description can be literally transferred to the quantum computer. The gates in this case are the unitary transformations, which concern only a few register qubits.

A *program* for such a computer is a sequence of gate operations. An additional "quantization" could be conceivable, through which the program itself would exist as quatum information. However, this is usually not meant when talking about programs of a quantum computer. The program sequence is simply fixed in a classical list.

The *input* for a quantum computer is made by the preparation of the registers. A part of this procedure is the *initialization*, i. e. the preparation of a defined initial state. The *output* is obtained by measurement at suitable registers. The result of a calculation is therefore always a matter of chance: two runs of a quantum computer will, as a rule, give different results. Only the probabilities of the results can be given in a meaningful way. This is also true for classical algorithms with built-in chance components. We are going to discuss later in which way useful results can be achieved in spite of this chance nature.

Building a quantum computer is nowadays the greatest challenge for experimental physics. The coherent manipulation of only a few qubits can already be regarded as a great success. Interesting programs for such computers already exist. Among these are two outstanding examples, both solving a classical problem more efficiently than any (known) classical algorithm. These are the Grover algorithm for the search of a marked element in a list (or a needle in a haystack) and the Shor algorithm for the factorization of large numbers. We are going to discuss the Shor algorithm in detail in Section 7.5, but first we want to clarify what we mean by the "efficient" solution of a problem.

Complexity

What is a difficult calculation problem? The mathematical theory dealing with this question is called complexity theory. This is not about the difficulty of the multiplication problem "137×42", but the difficulty of the problem "multiplication". As a measure of the difficulty, the number of steps needed to get to the solution is considered, which is roughly the time necessary for the solution.

Let us take as the simplest example the comparison of two types of problems, both already known from elementary school. The first problem is "count up to $N!$", where N is a place marker for a number that marks the special problem of this type. The second type of problem is: "write down the number $N!$". In the first year of school both problems have about the same difficulty, because N will not be larger than 10. This changes dramatically when the number N is chosen to be larger. It is already quite boring to count up to 1000, while writing down the four digits of "1000" goes very rapidly. When N is increased tenfold, the counting problem really takes ten times longer, while the writing problem only gets more difficult by the writing of just one more digit. The complexity theory is interested exactly in this type of differences. This theory asks about a given type of problem: how fast will the number of necessary calculation steps grow, when the initial data (here the number N) are increased? In this case the counting problem belongs to the class of problems with exponential effort, which means that the number of steps grows as a^n, where n is the number of input digits and a is a suitable number (here 10). In contrast to this, the writing problem belongs to the class of problems with polynomial effort: the number of steps grows as $p(n)$, where p is a polynomial function and n again the number of input digits (here $p(n) = n$).

Most of the other problems from elementary school belong to the polynomial class: the addition, the multiplication (two n-digit numbers, according to the method taught there, require about n^2 applications of the multiplication table up to ten, and the addition of n n-digit numbers also takes $p(n) \approx n^2$ steps). Also, the division, the determination of the largest common factor (following the Euclidean algorithm), and the calculation of transcendental functions like exp, sin, etc., belong to this class.

On the other hand, a member of the difficult, exponentially growing problems is the factorization of numbers, i.e. the determination of two integer numbers $P, Q > 1$, so that $P \times Q = N$ results in the given number. There is a simple method to solve this, namely the trial of all numbers up to $\sqrt{N}$, which is sufficient since both factors cannot be larger than $\sqrt{N}$. About a^n steps with $a = \sqrt{10}$ are necessary for an n-digit number. Modern classical factorization algorithms are much better, but they still remain in the exponential class. Most of the researchers in this field tend towards the opinion that in fact no polynomial algorithm exists, even though most likely nobody will succeed to prove this statement about *all possible programs* in the near future.

The problem of factorization is of the highest practical relevance. The standard encoding in public key systems on the Internet, the RSA algorithm, is based exactly on the difficulty of this problem. The number N is needed for the encoding, but for the decoding one also needs the factorization. So I can publish the key N for all data directed to me, and I am sure that nobody except myself can read the as-encoded data *as long as* nobody can succeed in solving the factorization problem. Usually, key numbers of 128 or 168 bit length are used, which corresponds to decimal numbers of about 50 digits. Even with great advances in computer technology, the factorization problem can be made more difficult by a certain factor with the simple enlargement of the key. With this, the race between encoding and decoding experts finally appears to be over in favor of the encoding experts (Singh (2000)).

The news about a factorization algorithm in polynomial time, the Shor algorithm, came as a real bombshell in this situation in the year 1995. The algorithm only had one minor flaw: it required a quantum computer and that was something people had just started dreaming about.

Problems quantum computers cannot solve

One argument against the idea that quantum computers can really achieve something new is obvious: quantum mechanics is defined by classical mathematical structures (matrices, partial differential equations, etc.) that can be processed on a classical computer at any precision. One only needs to write the given quantum algorithm in such terms and solve the corresponding equations, in order to obtain the probability distribution of the results of a quantum calculation as classical values.

The argument really shows that classically *unsolvable* problems also remain unsolvable for the quantum computer. An example in this respect is the word-problem of group theory: can two given expressions in the generators of a group be transformed into each other with the use of given algebraic rules (where interim results of any length can occur)? For this, it has been proven that no algorithm exists that could come to a decision in any case, and this also holds for a quantum computer.

Calculation of classical functions

The design of algorithms for quantum computers would really be difficult, when all the algorithms have to be reinvented. Fortunately all classical algorithms can be transferred to quantum computers (with small modifications).

For this purpose we recall the fact that quantum registers are distinct from classical registers by the additional possibility of superpositions. However, we can also ignore this additional possibility: every qubit is then always kept in one of the pure states HORIZONTAL or VERTICAL, the so-called *calculation basis*. In any case, where a classical program now requires a certain transformation of switching states, we also apply the corresponding transformation to the quantum register, in which the transformation of the basis states is exactly ("linearly") continued over all states by the superposition principle. With this it is guaranteed that a classical initial state (i. e. a state in which every qubit was prepared in one of the two possible states of the calculation basis) keeps this attribute and the state evolves exactly as in a classical computer.

However, there is a snag: not every transformation of the classical switching states can be extended to an allowed quantum-mechanical transformation (a "unitary operator"), because for this, the transformation needs to be *reversible*, i. e. it must be possible to reconstruct the input from the output unambiguously. Many classical gate operations do not meet this condition. For example "A OR B" is TRUE in three different cases, namely for (A, B) = (TRUE, TRUE), (TRUE, FALSE), and (FALSE, TRUE). It is impossible to find out afterwards which one of the three cases was the input of the operation. We can still construct the reversibility artificially, simply by memorizing the input, and thus instead of calculating $x \mapsto f(x)$, we calculate the function $x \mapsto (x, f(x))$ (Fig. 7.8). It is nevertheless not recommended to apply this procedure for the creation of a reversible process to every single step in a classical algorithm, because the required memory would explode exponentially with the number of steps. When it is applied a little smarter, i.e. with the removal of "junk data" in between, the procedure leads to the goal without an exorbitant additional expenditure also for general algorithms.[13]

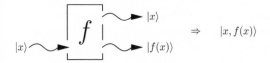

Fig. 7.8 Reversible calculation of the function f.

7.5 The Shor algorithm

7.6 Reduction to the problem of searching the period

The factorization of large numbers is not directly carried out in the Shor algorithm, but on a detour over another problem also known to be difficult, namely the problem to determine the period of the function $x \mapsto a^x \bmod N$. In this section we are first going to show that the solution of this problem also solves the factorization problem. In the following, N will always be the number to be factorized, and a is at first any chosen number that is sufficiently large, not a factor of N and an integer.

The idea is simply the binomial formula

$$a^{2m} - 1 = (a^m - 1)(a^m + 1) \tag{7.1}$$

Now we are searching for values of m that will give a number that can be divided by N on the left side, and consequently also on the right side. We then have the product of two numbers $N_\pm = (a^m \pm 1)$ on the right side, which can be divided by N. The same holds when we replace $N_\pm$ with the remainders from the division by N. At least one of these numbers must therefore have a true factor with N in

---

13) The procedure is older than the quantum computers and was originally developed with the idea that reversibly operating chips should consume less power. Some important works on this topic are from C. H. Bennett, who was also involved in launching the idea of the quantum computer.

common, which we can afterwards determine through the (very fast) Euclidean algorithm.

So now we are left with the problem to find an m so that a^{2m}, when divided by N, gives the remainder 1. For this, we can gradually calculate the powers a^n up to multiples of N and wait until within this sequence one value repeats itself. Namely, when we have $a^{n+p} \equiv a^n \bmod N$ (therefore both sides give the same remainder for the division by N), either we get $a^n \equiv 0 \bmod N$, which would contradict the choice of a, or we get $a^p \equiv 1 \bmod N$. Since maximum N different remainders for the division by N are possible, the sequence $a^n \bmod N$ *must* repeat itself at some point, which guarantees that the equation $a^p \equiv 1 \bmod N$ can be solved. In the case that p now turns out to be even, we simply set $m = p/2$ and get a product of two integer numbers according to Eq. (7.1), containing all factors of N. With a bit of luck, not all division factors lie in the same factor, i. e. it is not already the case that one of the factors can be divided by N, and the other one leaves the remainder ± 1. By using the fast Euclidean algorithm, we can then determine a nontrivial common factor of N and the first factor in Eq. (7.1), and thus have a true factor of N.

In the interesting applications, N could be a 100-digit decimal number. This would still not make the calculation with the remainders $\bmod N$ difficult, but of course it completely rules out the idea of testing 50-digit square roots of N as factors. The test of one billion factors per second, even parallel on one billion PCs, still takes a few million ages of the world. Neither is the trial of possible periods $a^p \equiv 1 \bmod N$ really practicable. So with complete justification one may ask, whether anything was gained by reformulating the problem. In fact, the reformulation would be nonsense as long as only classical computers are to be used again. So before we get to the details of how the Shor algorithm finds the necessary periods, we want to give a quick number example that shows the reformulation.

We take the factorization of the number $N = 21$ as an example and choose $a = 2$. Since we have $2^6 = 64 = 3 \times 21 + 1$, we find the period $p = 6$, thus $m = 3$ and with that $N_- = 2^3 - 1 = 7$ and $N_+ = 9$. Both numbers have a nontrivial factor in common with N. We could have had bad luck with the choice of a. With $a = 4$ for example, we get because of $4^3 = 64 \equiv 1 \bmod 21$ the odd period $p = 3$, and $a = 5$ gives $N_- = 124 \equiv 1 \bmod N$ and $N_+ = 126$, which can already be divided by N. The algorithm is summarized once again in a diagram in Fig. 7.9.

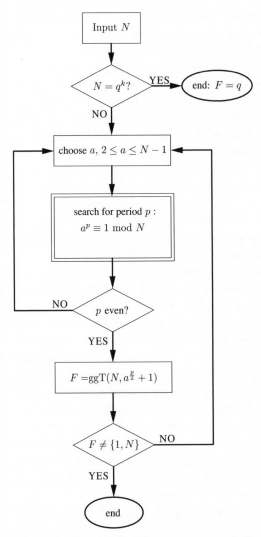

Fig. 7.9 The Shor algorithm. The quantum computer is only used in the double-framed part of the search for the period.

Stochastic aspects

A basic aspect of quantum computers is the fact that even for the same initial data, the results occur only with a certain probability. Now what kind of use do we have for a statement of the type "this is probably one of the factors we were looking for"?

For the factorization problem the answer is simple: it is very easy to check the result. We check the suggested factor rapidly, and when the test turns out negative, we let the computer run once more. Let us assume that a very fast computer would give us a factor with 50 % probability. A correct result after n runs would then be as sure as getting "head" at least once from tossing the coin n times. Even many small hit probabilities, of the order of 1 %, are still quite good. It would then simply take a few more trials.

This is not a specific problem of quantum computers. Also, classical algorithms very often contain stochastic elements. We already got to know an example, namely the classical part of the Shor algorithm. We assume that the search for the periods is carried out with certainty and immediately by a magic black box (algorithm theorists call such a box an oracle). We have seen that it is not yet sure that we are going to get a result, because we could have had bad luck with the choice of a, and found an odd period. When we throw the dice for a, the following question arises: how likely are we actually going to get an even period? The whole algorithm would be worth nothing without a clear estimate for this probability.

This estimate, which is a pure problem of number theory and has nothing to do with quantum mechanics, is one of the most important achievements when constructing an algorithm. We choose an odd N and ask for the probability W that an a, which is not a division factor, gives an even period p for which the factorization gives a true factor of N. This quantity is estimated as (Ekert and Jozsa (1996))

$$W \geq 1 - \frac{1}{2^{k-1}} \tag{7.2}$$

where k is the number of *different* prime factors in N. This estimate is the guarantee we were looking for, at least as long as $k > 1$. For $k = 1$, i.e. when $N = p^r$ with a prime number p, we have bad luck: not only does the estimate $W \geq 0$ become trivial, the algorithm actually fails miserably ($W = 0$ can occur). Since we cannot know in advance whether N is of this form, the factorization method would have failed anyway, had there not been fast (polynomial) factorization methods especially for the case of prime numbers available, which Shor also knew about. The description in Section 5.1 was therefore not correct yet. The complete Shor algorithm begins with the tests "N even?" and "$N = p^r$?", which can both be worked out in a purely classical way. Only then is the die thrown for an a and the quantum computer started for the search of the period.

We have discussed these subtleties of the Shor algorithm in so much detail in order to explain how important the "proof of effectiveness" of an algorithm with stochastic elements really is. This is especially true for the central part of the Shor algorithm, the period searcher we are now going to deal with.

The quantum-algorithm

Finding the period of a function is the problem that is solved with the Shor algorithm. This is applied to the function $f(x) = a^x \bmod N$, but practically any other function is equally suitable. The function should of course be coherently calculable.

Since above all we are interested in the runtime, we also demand something about the runtime behavior of the calculation of f. In the way that is typical for complexity theory, we consider problems characterized by a size parameter $L = $ length of the input string. For the factorization, this is $L = \log_2 N$. We now demand that a polynomial function $p(L)$ exists, so that $f(x)$ can be calculated for each given value x of an L-bit-register within a maximum number of $p(L)$ steps.

For the power function, one should therefore not make the mistake of calculating $a^x = a \cdot a \ldots a$, because that would mean up to 2^L multiplications. It is more practical to prepare the powers $a^r \bmod N$ for $r = 2^k$ by a successive calculation of squares, and to represent for a given x the power $a^x \bmod N$ as a product of all a^r that are shown by the binary factorization of x. Thus, one writes for example

$$a^{42} \equiv a^{32} \cdot a^8 \cdot a^2 \tag{7.3}$$

The number of square operations, as well as the number of factors necessary for a given x, then grow approximately proportional to L.

Step 1: The initial state

In order to bring our demand for the coherent calculability into play again, we generate at first a "register covering" superposition of the possible inputs, hence the quantum state

$$\Phi_1 = \frac{1}{\sqrt{2^N}} \sum_{x=0}^{2^{L}-1} |x\rangle \qquad (7.4)$$

Since the sum contains 2^L terms, it is once again not evident that this state can be produced in polynomial time. However, this is only due to the awkward formulation, because in fact we have a product state, in which each single register bit has to be put into the state $\frac{1}{\sqrt{2}}(|0\rangle + |1\rangle)$. Therefore, only L preparation operations are necessary.

Step 2: The coherent calculation of *f*

We now apply the coherent calculation of f to this initial state, which results in the state

$$\Phi_2 = \frac{1}{\sqrt{2^N}} \sum_{x=0}^{2^{L}-1} |x, f(x)\rangle \qquad (7.5)$$

For this, we have only used as many steps as necessary for the calculation of f on registers of this length, namely $p(L)$. Please note that this is the time for one single classical calculation. On the other hand, the state contains the information about *all* functional values of f! This effect, also called *quantum parallelism*, is one of the secrets of the success of the Shor algorithm.

This idea by itself, however, is still quite useless. Well, we could try to use the list of all functional values that is coded by Φ_2 as a reference table. We would then measure the two registers in the calculation basis, find a value for x and in the second register the corresponding functional value. However, the probability to hit exactly the value of x we are interested in is only 2^{-L}, and thus practically zero for large L.

The situation looks different when we are interested in a property of f that depends on all functional values together, and not in a very pronounced way on one single value. Periodicity is of this type, so we are going to apply a procedure that is used to extract a strong signal from weak periodic structures also in other fields of physics: the Fourier transformation.

Step 3: The quantum Fourier transformation

The Fourier transformation is a procedure that analyses any signal for its frequencies. The result of the Fourier transformation of a periodic signal contains as frequencies only the inverse period and its multiples, i. e. the "overtones". Even in the case of numerical approximations when considering functions that do not range to infinity and thus are not strictly periodic, or when no high frequencies can be resolved, is the Fourier transformation still essentially correct. From an approximately periodic function, it will generate a transformed function showing sharp maxima at the basic frequency and its multiples.

Fast ways to calculate the Fourier transformation have been investigated for a long time, because of its great importance for all physics. The so-called "fast Fourier transformation" (FFT) that transforms functions for n discrete values in $n \log n$ steps is quite famous. The quantum Fourier transformation, however, is much faster, because it uses just a few more than $\log n$ steps. The reason is that, similar to the preparation of the initial state (Step 1), the transformation can be broken down into operations that only concern single qubits. In the classical FFT algorithm, the counterpart of one single quantum operation of this kind is a step that has to be carried out for all possible values of the other bits. This takes a factor of n (total number of possible values) in the runtime behavior of the classical algorithm.

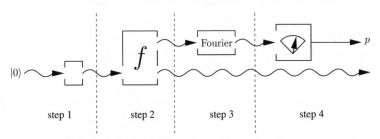

Fig. 7.10 The central part of the Shor algorithm for the search of the period of a function f.

Step 4: The measurement of the period

The Fourier transformation generates a strong constructive interference at the period frequency of the function f and its multiples. This simply corresponds to an increased probability of obtaining one of these values from the measurement of the register qubits. The details (Ekert and Jozsa (1996)) are again quite subtle: the maxima are not absolutely sharp and with a certain probability one can be quite wrong. This effect has to be estimated in order to make sure that the correct value is obtained with a sufficiently high probability. All these uncertainties are cured with the following last step.

Step 5: The verification

This is a purely classical step. With the assumed period p in the factorization example, we calculate the remainder of a^p divided by N. This goes within polynomial time, as described in the beginning of this section. When we get the value 1, the search for the period was successful and we continue in the classical part of the Shor algorithm with the question whether p is even and gives a factorization. In all the other cases, we have to go back to step 1 and possibly start with a new value a.

7.7 Realizations and outlook

I hope that I have explained the principles a quantum computer is based on in a fairly conclusive way: these are the basic principles of quantum mechanics, one of the best-confirmed physical theories available to us. Now is the whole thing really going to work? Is the gain due to the use of quantum principles not possibly lost somewhere else, so that the "law of the conservation of difficulty" holds again?

The critical points where such losses could occur are the tolerances of the components. The algorithm requires all the necessary operations to be carried out exactly. From this point, there is no difference for the designer of a quantum algorithm or the designer of classical algorithms. So what if a small error slips into each operation? Is it possible that these errors build up gradually, so that the exponential acceleration of the algorithm could finally only work with exponentially precise single operations (that definitely cannot be realized)?

This suspicion has already been expressed quite early, and up to now it cannot be simply denied. The entangled states are in any case more susceptible to disturbances than classical correlations and on the other hand some of the current ideas for the construction of error-tolerant classical computers fail. An essential principle for the prevention of error accumulation in classical computers is the *digitalization*: single bits are realized by circuits, which have two possible stable states. Small deviations are "slowed down" electronically, which means they are brought back into the next stable state under generation of heat. This mechanism functions essentially separately for each circuit. The space of quantum states (Fig. 7.7), in contrast to this, has no "engaging points"; only to define them would mean to get back to the classical data, and thus we would lose all possible advantages of the quantum computer right from the beginning.

Another idea for classical error control that unfortunately cannot work is the limitation of transmission errors with redundant sending. In this case every bit is sent for example three times, and at the end the result is determined according to the majority of the arriving bits. When every hundredth bit arrives falsely for the single transmission, two errors must come together in the case of the threefold transmission in order to generate a final error, which happens only approximately three times out of 10 000 transmissions. Exactly this procedure, however, does not work at all with quantum mechanics, because of the no-cloning theorem.

So what can be done in order to deal with the inevitably occurring errors? The second problem mentioned here can definitely be regarded as being solved: error-correcting codes have meanwhile been developed for quantum information. Without the copying step, quantum systems are distributed over several systems and thus secured against errors on one or a few subsystems. Instead of a threefold redundancy, as in the classical case, a fivefold redundancy is necessary, and errors of the order of magnitude ε can thus actually be suppressed to the order of magnitude ε^2. The application of such correction procedures has also been investigated. The critical question is in this case, whether the Shor algorithm can still get along with a polynomial growth of resources for finite tolerances of the components. It is difficult to give a simple answer to this question, because assumptions about the concrete realization have to be included in the description of the occurring results. A positive answer becomes apparent there, but the demanded tolerances are very utopian in view of the state of the art technology.

What physical systems should be considered for quantum computers?[14] We cannot give any answer that would also hold for the future due to the universality of quantum mechanics. A whole series of model systems already exists nowadays, which are useful at least for the demonstration of basic principles. These are — without claiming completeness or judging their relative contributions — the following:

- *Ion traps,* in which a few atoms are trapped inside an alternating electromagnetic field and in which the excited states of the stored ions serve as "qubits".

- *Nuclear spin resonance.* This is an established analytic method in chemistry. The qubits in this case are the spins of nuclei of different atoms in a molecule that is brought into a strong magnetic field and manipulated with microwave pulses. The signal of a single molecule, however, is so weak that one has to read the accumulated signal from a macroscopic amount of molecules. The problem is that the atoms must be addressed with different frequencies and therefore the maximum number of different qubits is very limited.

- *Electromagnetic cavity radiation.* The qubits are the occupational states of the resonant oscillations of a field in a — in most cases tiny — mirror-coated cavity. These systems are interesting as switches, where an atom flying through the cavity exchanges a photon with the field, which again is passed onto a following atom. In this way, an entanglement of the atoms becomes possible.

- *Optical gratings.* Similar to an ion trap, neutral atoms are kept here in a standing laser field. When putting the laser out of tune, the atoms can be made to interact by shifting them depending on their state. The interesting aspect of this idea (Raussendorf and Briegel (2001)) is that the calculation can

---

14) Compare to Sections 4.6 and 6.3.

be carried out in a way that a highly entangled many-particle state is generated just in the beginning and afterwards only operations like measurements and rotations at single particles are needed. The entanglement of the initial state is then "used up" gradually. For this type of quantum computer, one has a completely different view of the question of what should be considered as an "elementary step". From this, different complexity judgements can follow compared to the ones described in Section 7.4 for the gate structure.

– *Quantum dots* are structures that can be produced on solid-state objects with modern methods of chip production. Single electrons can be stored and manipulated in these structures. Some physicists believe that such structures offer the best chances for the construction of quantum computers with an interesting size due to the fantastic freedom with the design of wafer structures. The problem in this case is that the control of interactions with "undesirable" degrees of freedom is particularly difficult in the solid state. Therefore, even the realization of one qubit that can be freely manipulated over some longer time would be seen as a great success at the moment.

– *Atom chips.* This is an attempt to make good use of the possibilities of design in the chip technologies in order to keep, guide, and manipulate single atoms over the solid-state surface with specifically generated structures of electric and magnetic fields.

None of these experimental approaches is presently getting in any way close to a quantum computer that could solve any nontrivial problem. This present situation transferred to the world of classical computers corresponds to a stage at which the basic construction principles for computers are clear and one can already dream about a universal supercomputer. However on the practical side, only the first switching elements like levers, water valves, and electric relays are being developed.

Should the research aiming for a quantum computer be given up then? I don't think so, because it has already evolved into a research program with much wider objectives. It could roughly be described as the art of "coherent manipulation of quantum systems", which brings together experimental disciplines of a very different kind. The quantum computer is the long-term goal here that helps to focus the efforts. Whether a quantum computer of the Shor class will ever be built or not: we are going to learn a lot about the coherent manipulation of quantum systems through the research program in this direction and this will certainly open up a wide field of applications.

Besides this, little can be promised about a quantum computer being built that will seriously compete with classical computers in 50 years from now nor can one exclude that this is already going to happen in 10 years. Such prognoses depend mainly on thrusts of ideas in the art of performing experiments as well as on their theoretical understanding. And if these could be predicted, science would be only half as interesting as it is.

7.8 Bibliography

A lot of good literature on the topic of quantum information processing exists meanwhile. Four volumes in which the experts in the field present different aspects, and that are therefore very useful for an overview are

- H.-K. Lo, S. Popescu, T. Spiller (1998) (eds.): *Introduction to Quantum Computation and Information,* World Scientific, Singapore.

- G. Alber, T. Beth, M. Horodecki, P. Horodecki, R. Horodecki, M. Rötteler, H. Weinfurter, R. F. Werner, and A. Zeilinger (2001): *Quantum Information — An introduction to Basic Theoretical Concepts and Experiments,* Springer Tracts in Modern Physics, volume 173.

- D. Bouwmester, A. K. Ekert and A. Zeilinger (2000): *The Physics of Quantum Information,* Springer.

- M. A. Nielsen and I. L. Chuang (2001): *Quantum Computation and Quantum Information,* Cambridge Univ. Press. A good textbook for the whole field.

A useful source are the Internet pages of active research groups on which up to date literature references can be found, hopefully also in the future. A good starting point is the page from the Braunschweig group: http://www.imaph.tu-bs.de/qi.

References

- D. Bohm, *Phys. Rev.* **85** (1952) 166 und 180.
- D. Deutsch (1997), *The fabric of reality,* Penguin.
- A. Ekert and R. Jozsa (1996): "Quantum computation and Shor's factoring algorithm", *Rev. Mod. Phys.* **68**, 733–753.
- R. P. Feynman (1982): "Simulating physics with computers", *Int. J. Theor. Phys.* **21**, 467.
- E. Nelson, *Phys. Rev.* **150** (1966) 1079.
- R. Raussendorf and H.-J. Briegel (2001): "A one-way quantum computer", *Phys. Rev. Lett.* **86**, 5188.
- C. E. Shannon (1948): "A mathematical theory of communication", *Bell System Technical Journal* **27** (1948) 379–423 and 623–656.
- S. Singh (2000): *Geheime Botschaften,* Hanser, München.
- W. K. Wootters, W. H. Zurek (1982): "A single quantum cannot be cloned", *Nature* **299**, 802–803.

8

Decoherence and the transition from quantum physics to classical physics

Erich Joos

A consistent application of quantum theory shows that classical and quantum physics are connected with each other in a very different way from that described in the usual textbooks. In the last two decades, the phenomenon of decoherence has been recognized and investigated theoretically and experimentally and found to be an essential element of the transition from quantum physics to classical physics. Regarding the related problem of the interpretation of quantum theory, the conclusion can be drawn that from the multitude of solutions suggested in the course of time only two directions remain as being consistently feasible.

8.1 Classical and quantum physics

The relationship between classical and quantum physics has been in the center of the discussion about the interpretation of quantum mechanics since the beginnings of this theory. Does quantum theory contain classical physics as a limiting case, in a similar way as relativistic mechanics is reduced to the familiar Newton's theory when all velocities are small compared to the speed of light?[1] Or is classical physics still necessary to be able to pursue quantum physics at all? How do both theories fit together?

For a long time, the situation was overshadowed by the orthodoxy of the Copenhagen school. The interpretation of the 1920s, born out of despair, was extremely successful from a pragmatic point of view,

---

1) See also Section 1.4.

Entangled World, Jürgen Audretsch
Copyright © 2005 WILEY-VCH Verlag GmbH & Co. KGaA, Weinheim
ISBN: 3-527-40470-8

but the inner contradictions were felt as being painful and unsatisfying only by a few, among them the two prominent cofounders of quantum theory, Einstein and Schrödinger. The further the range of application of quantum theory (it evolved from Bohr's theory of the structure of the atom) was extended, the clearer it became that quantum mechanics describes fundamental properties of nature and is not just a preliminary construction, which would soon be replaced by a better theory anyway. In addition, experimental physicists succeed more and more to advance into critical intermediate domains where the difference between "microscopic" and "macroscopic" becomes questionable, which makes a uniform fundamental theory even more a necessity. From the present point of view this can only be accomplished by some sort of quantum theory. But what a picture of nature does this theory actually provide, and how does the classical world that is so familiar to us from our daily experience fit in?

Kinematics: How do I describe an object?

In order to become clearly aware of the problems arising when asking for the connection between classical physics and quantum theory, it is a good idea to take a step back and think about how a physical theory is typically constructed.

The first step for every description of nature is a conceptual scheme for the characterization of physical objects, let it be raindrops, the trees in my garden, the moons of Jupiter or electrons in a computer monitor. The mathematical terms that serve this purpose form the kinematical basis of the corresponding theory. So at first let us compare the mathematical description of objects in classical and quantum physics.

A simple example is a mass point (e. g. the center of gravity of a dust particle). In classical physics, a complete description requires specification of the position x and the momentum (velocity) p. The quantum-mechanical description looks completely different. Instead of particle coordinates, wave functions are used, which do not refer to a single point in space, but are extended over some region. Most importantly, these wavefunctions can be superimposed: from two states Φ and Ψ a new state can be obtained as a sum, $\Psi + \Phi$, which shows in general a completely different behavior from Φ or Ψ. All the "strange" features of quantum objects can in the end be reduced to the fact that quantum objects obey a superposition principle.

Table 8.1 Comparison of the basic kinematic concepts of classical physics and quantum theory. The connection between the left and the right side is not obvious at all.

classical physics	quantum theory
Galilei, Newton, ...	**Schrödinger**
mass point: $(\vec{x}, \vec{p})$	wave function: $\Psi(\vec{x})$
fields: $\vec{E}(\vec{x})$, $\vec{B}(\vec{x})$	superposition principle: $\Psi, \Phi \to \Psi + \Phi$

The classical analogy for this superposition are wave phenomena found for water waves or radio waves. Already there, the interference effects that are typical of waves can be seen. The superposition principle in quantum theory, though, is more general, because it holds for any kind of physical state. An important consequence of this is the "entanglement", which arises when states of composite systems are superimposed. It is possible, for example, to superimpose states with different numbers of particles (photons), which is essential for a correct description of laser light. Now, what do particles have to do with waves? This is an old question. At least for macroscopic objects we are quite sure that a description as particles (that is, spatially localized objects) or fields *in space* is very good, whereas in the microscopic domain such a picture turned out to be untenable.

Dynamics: How does an object evolve in the course of time?

The second part of the theory must now specify how the considered objects change in the course of time. As before: the laws of motion in classical and in quantum physics look completely different.

In classical mechanics mass points move following Newton's law $\vec{F} = m\vec{a}$, or, expressed in a more formal way, they move in accordance with Hamilton's equations (see Table 8.2). This fundamental law of motion defines the evolution of position and momentum in time. An analogous law of motion in quantum theory is Schrödinger's equation, which determines the evolution of wave functions in time. But here we come across a very strange problem: the Schrödinger equation does not seem to hold in just any case, because a second law is used for measurements, which replaces one wave function by another with a certain probability (the so-called "collapse of the wave function"). Such a theory is strictly speaking inconsistent because it has *two* laws of motion. However, every physicist knows when one law or the other has to be applied: Schrödinger's equation only holds for isolated systems, the collapse has to be used when a measurement is performed. But what is a measurement?

Table 8.2 Comparison of the temporal evolution of physical states in classical physics and in quantum theory.

classical physics	quantum theory
Newton/Hamilton	**two (!) laws of motion**
$\vec{F} = m\vec{a}$	1. Schrödinger equation
or	$i\hbar \dfrac{d}{dt}\Psi = H\Psi$
$\dot{x}_i = \dfrac{\partial H}{\partial p_i}$	2. collapse of the wavefunction
$\dot{p}_i = -\dfrac{\partial H}{\partial x_i}$	$\Psi = \displaystyle\sum_n c_n \Psi_n \rightarrow \Psi_k$
Maxwell equations	with probability $|c_k|^2$
for $\vec{E}(\vec{x})$, $\vec{B}(\vec{x})$	

Obviously there are serious difficulties in recognizing a clear connection between classical and quantum physics at a first glance. Therefore there is no surprise that again and again the same questions are discussed:[2]

- What is the meaning of the wave function Ψ?

- What is the meaning of the collapse of the wave function?

and first,

- What is the precise relationship between quantum physics and classical physics?

2) See also Section 1.10.

For more than 75 years now, we have had at our disposal an extremely successful quantum theory as a fundamental description of nature. So what could be more obvious to do than to look for the answer to all those questions in the established works? Let us open any textbook on quantum theory. What do we find there?

8.2 The fairy tale about the classical limit

A famous series of textbooks on theoretical physics comes from Landau and Lifschitz. The volume on quantum mechanics contains the following statements:

> Quantum mechanics also contains classical mechanics as a limiting case.

> ... The transition from quantum mechanics to classical mechanics is analogous to the transition from wave optics to geometrical optics.

> ... Then one can state that the wave packet is moving along the classical trajectory of a particle in space.

Formulations like the above can be found in nearly all textbooks on quantum mechanics. The explanation given for this quantum–classical correspondence is, for example, that the well-known spreading of the wave packet has no relevance for objects with macroscopic mass, because it would take an incredibly long time. Besides this, some mathematical theorems like the one from Ehrenfest come to help, which show that mean values in quantum theory approximately follow the classical trajectories. For the calculation of molecules, approximation procedures for the solution of Schrödinger's equation (going back to Born and Oppenheimer) are used, which result in quite classically looking states. That is, molecular wave functions are constructed, where atoms or groups of atoms show a well defined spatial arrangement just like the pictures known from chemistry books. Note that all these approximation procedures start from Schrödinger's equation. So everything appears to be all right, or perhaps not?

Criticism I

A very essential feature of quantum theory is the superposition principle, as mentioned before: *new* states can be constructed by linear combination (superposition) of two arbitrary states of a system. So when Ψ_1 and Ψ_2 are two solutions of Schrödinger's equation, *any* state of the form

$$\Psi = c_1\Psi_1 + c_2\Psi_2$$

for just any (complex) coefficients c_1 and c_2 (apart from normalization) is also a solution. Obviously, there exists a great number of them. The big problem now is this: even though Ψ_1 and Ψ_2 may describe "classical states" (like spatially localized objects in different places or dead or alive cats), most of the combinations above are totally "nonclassical". But this means: *almost all of the states, which are (via the superposition principle) allowed by quantum theory, are never observed!* We only see either *dead* or *living* cats and we cannot even imagine what such a strange intermediate state should actually mean.[3]

Perhaps one could talk our way out of this with the idea that such states just never come into being and that the dear Lord has created all macro-objects in the well-familiar classical states. Unfortunately, this will not take us any further, since exactly during a measurement process a microscopic superposition is amplified towards macroscopic dimensions, so that the occurrence of these states becomes *inevitable* (exactly this was the essence of Schrödinger's cat argument from 1935).

The above argument is not new: it can also be found in the correspondence of Max Born with Albert Einstein. I quote from a letter of Einstein to Born dated January 1st, 1954:

> Your opinion is quite untenable. It is incompatible with the principles of quantum theory to require that the Ψ-function of a "macro"-system should be "narrow" with respect to macro-coordinates and -momenta. Such a demand is incompatible with the superposition principle for Ψ-functions.

---

3) The often-expressed idea that the cat is either dead or alive, just that it is not known, turned out to be untenable. Superpositions show *completely different* properties from their components.

He adds as an explanation in a footnote:

> Let Ψ_1 and Ψ_2 be two solutions of the same Schrödinger equation. Then $\Psi = \Psi_1 + \Psi_2$ is a solution of the Schrödinger equation as well, with the same claim of being a description for a possible real state. When the system is a macro-system and Ψ_1 and Ψ_2 are "narrow" with respect to the macro-coordinates, this is not the case anymore in the vast majority of the possible cases for Ψ. Narrowness regarding the macro-coordinates is a demand that is not only *independent* from the principles of quantum mechanics, but also *incompatible* with these principles.

Any attempt to justify classical physics from a quantum theory must explain first, why we observe only very *special* states (which are then *called* "classical"). The occurrence of only special states is by the way not just found in domains, which are quite obviously "macroscopic". Even a single sugar molecule is always observed as being right or left handed, but never in a superposition state, which is otherwise typical in the atomic domain. The term "macroscopic" therefore has nothing to do with "size" in the first place. There are also well-known (and well-understood) "macroscopic quantum states", for example superconductors.

Criticism II

The second important point of criticism is based on a quite elementary observation, yet with enormous consequences: macroscopic objects are never isolated because they are *always* interacting with their natural surroundings. In order to become convinced that this is true, one just needs to open one's eyes. We see the things around us only for the reason that they are scattering light. The scattering of light off an object implies on the one hand that it clearly may not be considered to be isolated, and on the other hand this means that an interaction occurs, which has properties similar to a measurement: The radiation field *after* the scattering at an object contains some "information" about the position of the object that was scattering. We know already that a measurement in quantum theory has special consequences. Among other things, the measurement destroys exactly the typical quantum behavior, namely the ability to interfere. Interferences can never occur, if such measurements happen all the time.

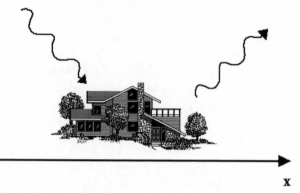

Fig. 8.1 Macroscopic objects are always "under observation" by their natural environment. They are scattering photons or molecules. Even in the vacuum of outer space, they cannot escape from this interaction. In this way, the position x becomes a classical quantity.

This would directly give an explanation for the fact that a superposition of a dead or alive cat or of other macro-objects at different places is never observed. When this interaction is really essential (a *quantitative* question that will be discussed later on), no Schrödinger equation can hold for macro-objects. The textbook derivations of the "classical limit" mentioned above are invalid also for this simple reason.

What kind of an equation instead of the Schrödinger equation holds then? This question can unfortunately not be answered generally in a simple way, because the consequences of the interaction of a system with its surroundings can be very diverse. So there is no other way but to take the law of nature assumed to be fundamental, namely the Schrödinger equation, and apply it to typical situations. The fact mentioned before that the surroundings act like a measuring instrument for macro-objects, is particularly interesting. Therefore it is appropriate to start with a look at the quantum theory of measurement.

Superpositions, interferences, density matrices

All predictions of quantum theory can be expressed in a compact form by using the so-called density matrices. They are constructed from wavefunctions, and the probabilities to find certain states upon measurements can be calculated with them. Let's take a look at a two-state system as an example. This is the simplest quantum-mechanical object and can be realized, for example, by polarization of an electron spin or of light. In some modern applications, the term "quantum-bit" is used. The most general state is a linear combination of two basis states[4] $|1\rangle$ and $|2\rangle$ of the form

$$|\Psi\rangle = a|1\rangle + b|2\rangle$$

The density matrix that corresponds to this state has *four* maxima (Fig. 8.2), two of them give simply the probability to find $|1\rangle$ or $|2\rangle$ (therefore they have the height $|a|^2$ or $|b|^2$, respectively). The additional peaks are responsible for all interference effects that may occur in appropriate situations.

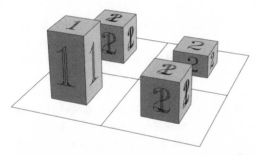

Fig. 8.2 Density matrix of a superposition of two basis states $|1\rangle$ and $|2\rangle$. The two components along the diagonal give the probability to find one of the two states. The occurrence of the two interference terms shows that *both* components are present and therefore the state is to be characterized by $|1\rangle$ *and* $|2\rangle$.

On the other hand, when an ensemble of $|1\rangle$ *or* $|2\rangle$ has to be described, that is (among other things) the situation *after* a measurement, the density matrix corresponds to a probability distribution

4) In the following I will mostly use Dirac's "ket" terminology for quantum states. This is particularly useful for general considerations, since then one can restrict oneself to a rough characterization (e. g. with "quantum numbers" or simply with words, as in $|$dead cat$\rangle$).

and therefore it has only two components, because members of an ensemble of course can not interfere with each other (see Fig. 8.3).

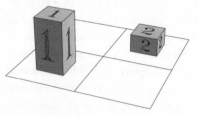

Fig. 8.3 Density matrix of an ensemble of two states $|1\rangle$ and $|2\rangle$. Since either $|1\rangle$ *or* $|2\rangle$ is present, no interference terms exist.

This corresponds to a black-box situation: a certain state is present, but nobody has had "a look at it" yet, therefore a probability distribution is an adequate description. Therefore, an important criterion for classical behavior of an object will be the check, whether or not interference terms can be found in the corresponding density matrix. Then one can at least say that the object *appears* in a way *as if* it is either in the state $|1\rangle$ or $|2\rangle$.

An analogy to the two situations described above is the interference of two radio stations. When both of them are on air simultaneously, their signals can superimpose coherently, so that (usually disturbing) interferences occur. This corresponds to the above wavefunction $|\Psi\rangle$ or the density matrix in Fig. 8.2. When only one of the two radio stations works (no matter which one), there are of course no interferences, corresponding to Fig. 8.3. Similar arguments can be applied to diffraction experiments at a double slit, which can show an interference pattern only when *both* partial waves (through each slit) are present, as in Fig. 8.2.

8.3 Quantum theory of the measurement process

From a physical point of view, a measuring instrument is nothing else but a special system whose state contains information about the "object of the measurement" after interacting with it: The state of the measuring instrument allows conclusions to be drawn about the state of the measured system. So let us apply the quantum theoretical formalism to such a situation in a way that was first described by Neumann in his book "Mathematische Grundlagen der Quanten-theorie" published in 1932.

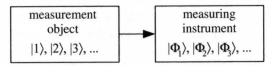

Fig. 8.4 Simple von Neumann model of the measurement process. Information about the state $|n\rangle$ of the object is transferred to the "pointer" that shows the value n after the measurement, corresponding to a quantum state $|\Phi_n\rangle$.

We assume that there exists a series of possible states $|1\rangle$, $|2\rangle$, ... $|n\rangle$ for the measured object, which are supposed to be distinguished by means of the measuring instrument. In the case of the so-called ideal measurement nothing happens to the object of the measurement, but the state of the measuring instrument changes from an initial status $|\Phi_0\rangle$ to $|\Phi_n\rangle$, for example[5]

$$|1\rangle\,|\Phi_0\rangle \xrightarrow{t} |1\rangle\,|\Phi_1\rangle$$

or

$$|2\rangle\,|\Phi_0\rangle \xrightarrow{t} |2\rangle\,|\Phi_2\rangle$$

depending on whether the object was initially in $|1\rangle$ or in $|2\rangle$. So far all this corresponds to the classical idea of a measurement. But now the superposition principle leads to a severe problem: For any initial state $a|1\rangle + b|2\rangle$, by adding the two expressions, we can directly write down the corresponding solution of the Schrödinger equation, namely

$$(a\,|1\rangle + b\,|2\rangle)\,|\Phi_0\rangle \xrightarrow{t} a\,|1\rangle\,|\Phi_1\rangle + b\,|2\rangle\,|\Phi_2\rangle$$

What we get, however, is not a single measurement result, but a superposition of all possible pointer positions (in each case jointly, "quantum correlated", with the corresponding object states). The fact that after this interaction both parts now have only a *common* wave-function is essential — this is an *entangled state*.

If the rules for measurement probabilities are now applied to the measured system (or the measuring apparatus) — for example by

5) Compare to Sections 2.3 and 5.1.

calculating the corresponding density matrices — one finds that no interferences occur any longer. This corresponds to the well-known fact that a measurement "destroys" interferences. However, the final state from above is still completely coherent, the interferences are actually still present, but they can only be found through measurements at the combined system. Therefore, interferences are hard to detect for "large" systems already for practical reasons, because one needs access to all relevant degrees of freedom.

The vanishing of interferences for subsystems — decoherence — is a direct consequence of quantum-mechanical nonlocality, which finds its expression in the "strange" attributes of entangled states. Since coherence (ability to interfere) is an attribute of the *overall* state, it apparently disappears for observations at subsystems, but is still existing.

We have pointed out above that especially macroscopic objects strongly interact with their natural surroundings. The quantum-mechanical description by means of the von Neumann model must therefore, as a consequence, also include the environment of the measurement device. This can easily be done — at least schematically.

The origin of classical properties

Since the surroundings of the measuring apparatus also work as a measuring instrument, we can also apply the above dynamics also to the interaction with the environment, and thus just extend the scheme of the previous section.

If the measuring instrument points at $|\Phi_1\rangle$, the surroundings $|U\rangle$ will recognize this very rapidly. Therefore, we have a chain of the type

$$|1\rangle\,|\Phi_0\rangle\,|U_0\rangle \to |1\rangle\,|\Phi_1\rangle\,|U_0\rangle \to |1\rangle\,|\Phi_1\rangle\,|U_1\rangle$$

where $|U_1\rangle$ means that the "rest of the world" has seen the pointer position $|\Phi_1\rangle$ of the (macroscopic) measuring instrument.

As a consequence, the quantum-mechanical description of the locally accessible system object of the measurement + measuring instrument changes: no interferences can be observed any longer, as expected. This is formally expressed in the way that the nondiagonal elements (interference terms) have disappeared from the density matrix as in Fig. 8.3. The macrosystem "measuring instrument" behaves therefore "classically" in this sense, *because it is measured as well*.

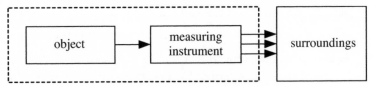

Fig. 8.5 Realistic extension of the von Neumann model. Information about the state of the object of the measurement is transferred to the measuring instrument. The pointer position is then registered extremely rapidly by the environment. Therefore, no interferences can be found at the local system object + measuring instrument. This "feedback" is a direct consequence of the quantum-mechanical nonlocality (entanglement), that is, the fact that the subsystem cannot have its own state and only a common entangled wave function exists.

But is this system now truly classical? We will get back to this delicate question at the end of this contribution. As far as it concerns the behavior of our systems — described by the corresponding density matrix — all results are independent of any preferred interpretation of quantum theory. For this reason, all physicists agree about the phenomenological relevance of decoherence, but not about its interpretation.

8.4 Examples

Spatial localization

All macroscopic objects always appear to be well localized in (conventional) *space* (but not in momentum space). This cannot be understood by considering these objects alone. Microscopic systems are normally observed in energy eigenstates. For a free mass point this would be an extended (plane) wave, just the opposite of a localized object.

In a realistic assessment of macroscopic systems it turns out that scattering processes are extraordinarily important (see Fig. 8.1). The state of the scattered particle (for example a photon or a molecule) depends on the position of the scattering center, the macroscopic object.

The scattered particle "measures" its position x. This can be formally described as a kind of a measurement-like interaction by

|object at position $x\rangle$|incoming particle$\rangle$

$\rightarrow$ |object at position $x\rangle$|particle scattered off position $x\rangle$

According to the quantum theory of the measurement process as outlined above, one can expect that interferences will be destroyed through such processes over large enough distances. More precisely, this holds for those distances that the surroundings can "distinguish". From optics it is known that the resolution of a microscope depends on the wavelength of the light source that is used. Here also, the large number of scattering processes is very important. The distance over which interferences still occur is usually expressed by the "coherence length" l. So one can expect that a double-slit experiment with a slit distance larger than l will not show interference patterns any longer.

From a simple model for the scattering process (see box "decoherence of a dust particle"), one can derive a decay of this coherence length according to

$$l(t) = \frac{1}{\sqrt{\Lambda t}}$$

For physicists: decoherence of a dust particle

To a first approximation recoil can be ignored, so that a single scattering process with the scattering center at position x follows the von Neumann scheme, thus

$$|x\rangle|\Phi_0\rangle \rightarrow |x\rangle|\Phi_x\rangle = |x\rangle S_x|\Phi_0\rangle$$

There, S_x is the scattering matrix with the scattering center at position x. For each single scattering process, the density matrix $\rho(x, x')$ that describes the position of the scattering center is multiplied by a factor that is given by the overlap of the corresponding scattering states $|\Phi_x\rangle$ and $|\Phi_{x'}\rangle$,

$$\rho(x, x') \rightarrow \rho(x, x') \left\langle \Phi_0 \left| S_x^{\dagger'} S_x \right| \Phi_0 \right\rangle$$

This factor is approximately zero when the wavelength of the scattering particle is smaller than the distance $|x - x'|$ (resolution of the microscope):

$$\left\langle \Phi_0 \left| S_x^{\dagger\prime} S_x \right| \Phi_0 \right\rangle = \begin{cases} 0 & \text{if } |x - x'| \gg \lambda \\ 1 - O[(x - x')^2] & \text{if } |x - x'| \ll \lambda \end{cases}$$

The case of low resolution in the second line is of particular interest. An effect of such "incomplete" measurements is not expected at first glance since a single scattering process cannot resolve the distance $|x - x'|$ and therefore cannot destroy the coherence between x and x'. But with a great number of such scattering processes an exponential damping of coherence results,

$$\rho(x, x', t) = \rho(x, x', 0) \exp[-\Lambda t (x - x')^2]$$

where the "localization rate" Λ is given by

$$\Lambda = \frac{k^2 \sigma_{\text{eff}} N v}{8 \pi^2 V}$$

Here, k is the wave number of the incoming particle, σ_{eff} is of the order of magnitude of the total scattering cross section and $N v / V$ is the flux density of the scattering particles.

	$a = 10^{-3}$ cm dust particle	$a = 10^{-5}$ cm dust particle	$a = 10^{-6}$ cm large molecule
cosmic background radiation	10^6	10^{-6}	10^{-12}
300-K photons	10^{19}	10^{12}	10^6
sun light	10^{21}	10^{17}	10^{13}
air molecules	10^{36}	10^{32}	10^{30}
laboratory vacuum	10^{23}	10^{19}	10^{17}

Localization rate Λ in $\text{cm}^{-2} \text{ s}^{-1}$ for differently sized "dust particles" and different scattering processes. This parameter shows how fast interferences between different positions are destroyed.

The scattering of air molecules is particularly efficient because of their small de Broglie wavelength.

The numerical value of Λ depends on the actual situation and it can be extraordinarily large. A dust particle with a radius of 10^{-5} cm in interaction with the cosmic background radiation for example gives $\Lambda = 10^{-6}$ cm^{-2} s^{-1}, and with thermal radiation at room temperature $\Lambda = 10^{12}$ cm^{-2} s^{-1}. The scattering of air molecules gives $\Lambda = 10^{32}$ cm^{-2} s^{-1} and even in the laboratory vacuum $\Lambda = 10^{19}$ cm^{-2} s^{-1} (10^6 particles per cm^3). This means that an initially present coherence over considerable distances is destroyed extremely rapidly. On the other hand, interferences are observed for small objects like electrons or neutrons. An interference experiment with fullerenes (C_{60} molecules) was even carried out recently. Whether an object still shows interference or not is therefore a *quantitative* question. Trying to get to the experimental limits in order to test quantum theory with objects as large as possible surely represents an important test of quantum theory.

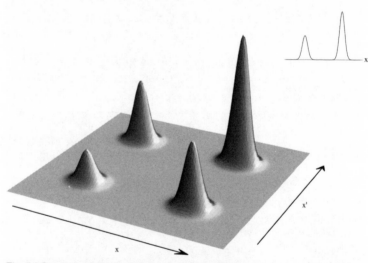

Fig. 8.6 Density matrix of a superposition of two different positions of an object. The nondiagonal terms show the presence of coherence between the two components of the wavefunction (upper right in the picture), analogous to Fig. 8.2.

A complete dynamical description of a mass point must include the inner dynamics of the system as well. The latter alone leads to the well-known spreading of the wave packet, described in all textbooks, and thus to an increase of the coherence length. The resulting equation includes, besides the Schrödinger evolution, additional terms taking care of scattering processes (see box "equations of motion of a dust particle").

For physicists: equations of motion of a dust particle

The temporal evolution of the density matrix of a dust particle can be described as a combination of the Schrödinger (von-Neumann) equation with additional terms describing the influence of the scattering processes,

$$i\dot{\rho} = [H_{\text{internal}}, \rho] + \frac{\partial \rho}{\partial t}\bigg|_{\text{scattering}}$$

The essential part for larger objects is the exponential decay of the coherence between different positions. For a "free" particle one has $H_{\text{internal}} = p^2/2m$ and for one spatial dimension, the equation of motion is explicitly written as ($\hbar = 1$)

$$i\frac{\partial \rho(x, x', t)}{\partial t} = \frac{1}{2m}\left[\frac{\partial^2}{\partial x'^2} - \frac{\partial^2}{\partial x^2}\right]\rho - i\Lambda(x - x')^2\rho$$

Extended models, which take recoil from the scattering into account ("quantum Brownian motion") give more complicated equations, for example

$$i\frac{\partial \rho(x, x', t)}{\partial t} = \left[\frac{1}{2m}\left(\frac{\partial^2}{\partial x'^2} - \frac{\partial^2}{\partial x^2}\right) - i\Lambda(x - x')^2 + \right.$$
$$\left. i\gamma(x - x')\left(\frac{\partial}{\partial x'} - \frac{\partial}{\partial x}\right)\right] \cdot \rho(x, x', t)$$

The difference between decoherence (Λ-term) and dissipation (γ-term) can be studied here. In the macroscopic domain, decoherence is far more effective than dissipation.

Analysis of the solutions of this equation shows that in all realistic situations the macro-object remains localized and the spreading of the wave packet that is discussed in textbooks does not occur. For long times, the measurement by scattering processes always dominates and the coherence length decreases. In a next step, the recoil during a scattering process can also be taken into account; the coherence length that finally results at equilibrium is the so-called thermal de Broglie wavelength $\lambda = h/\sqrt{(mk_BT)}$ of the object (see also the remarks about Brownian motion below). So the result of these considerations can be subsumed in the following statement:

> *All "macroscopic" objects are localized down to their thermal de Broglie wavelength.*

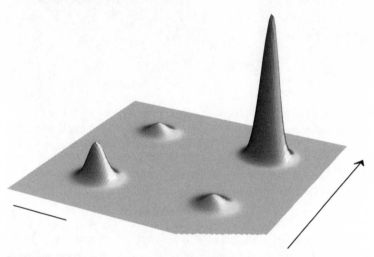

Fig. 8.7 The interaction with the surroundings damps the coherences.

This de Broglie wavelength is very small, for the dust particle mentioned above it is about 10^{-14} cm at room temperature. The high effective resolution of scattering processes is also why the spatial structure of molecules is well defined, except for very small molecules like ammonia, where a completely quantum-mechanical behavior is observed (maser). The pictures that are commonly used in chemistry are therefore indeed justified, but only for the reason that large molecules are not isolated systems and their spatial structure is continuously measured by their surroundings. From the point of view of quantum mechanics the question arises immediately, as to why we are able to assign to most molecules a well-defined spatial structure. Consider, for example, a pyramidal molecule of four atoms (or atomic groups). With the exception of very small molecules, such objects can successfully be described with spatially oriented models, in the way indicated in Fig. 8.8. In the case of four different ligands, this is an optically active molecule, which can be observed either in a right-handed or a left-handed configuration (a racemate is simply a *mixture* (not a superposition!) of both types).

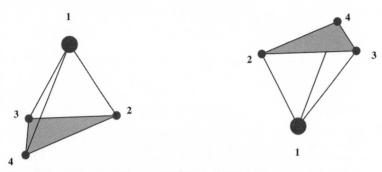

Fig. 8.8 Most molecules show a well-defined spatial arrangement of atoms or groups of atoms. Many optically active molecules occur in two mirror-image configurations. Quantum mechanically one would rather expect a superposition of these classical states. But this is never observed.

The transition between these two mirror-image configurations corresponds quantum mechanically to a tunnelling of atom 1 through the plane, a process that is formally described by the quantum theory of a double-bottomed potential (Fig. 8.9). For reasons of symmetry, the ground state is distributed over both minima. Such a state does not correspond to an actual spatial configuration, but is formally similar to Schrödinger's cat. The tunnelling probability is in most cases very small, which means that the molecule stays in one oriented configuration *as soon as* it is in one of the two minima. Before the advent of decoherence theory there were many attempts over decades to explain whatever distinguishes the classical states from any of their superpositions.

The scattering of photons, and first of all of other molecules, results also in this case in the destruction of interferences between the classical configurations, as discussed above. The fact that small molecules like ammonia, in contrast, behave totally quantum mechanically, can be understood in detail when the corresponding processes are quantitatively discussed and the different time scales are calculated.

Yet another point is worth mentioning here. The identification of the considered scattering processes with the well-known phe-

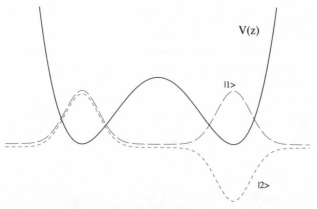

Fig. 8.9 The double-well potential gives a simple model for the tunnelling between the two classical configurations (wavefunction concentrated in one of the minima) of Fig. 8.8. The quantum-mechanical ground state $|1\rangle$ as well as the exited state $|2\rangle$ correspond to a wave function that is spread over *both* minima, and therefore does not match any geometrically defined configuration.

nomenon of Brownian motion comes naturally. Brownian motion is indeed a general consequence of scattering of many "small" objects at one "larger" object. However, decoherence is a typical quantum phenomenon and at first it has nothing to do with classical collision processes. The reason for this is the fact that decoherence is just a consequence of quantum-mechanical entanglement in the sense described in the beginning, but it does not include dynamical effects like for example recoil. This means that decoherence can occur without the collision changing the state of the object. This can also be found formally when studying the quantum-mechanical analog of Brownian motion.

The comparison of the corresponding equations shows that in the theory of Brownian motion an additional term is used (see box "equations of motion of a dust particle"), which contains the friction constant γ (in simple models γ is given by $\Lambda = m\gamma k_B T/\hbar^2$; here T is the temperature of the surroundings, k_B is the Boltzmann constant and m is mass of the object under consideration). When the decoherence term is compared with the friction term at a distance Δx, the following ratio is obtained:

$$\frac{\text{decoherence}}{\text{friction}} \approx \frac{m k_\mathrm{B} T}{\hbar^2} (\Delta x)^2 = \left(\frac{\Delta x}{\lambda_\mathrm{th}} \right)^2 \to 10^{40}$$

In this case T is again the temperature of the environment, m the mass of the object and λ_th the so-called thermal de Broglie wavelength. When typical macroscopic numbers are inserted in this equation, for example 1 g for the mass and 1 cm for the distance, a ratio of the order of 10^{40} results! It is thus obvious that quantum-mechanical decoherence is far more important than classical friction.

Experiments were performed recently that follow the onset of decoherence in the microscopic domain for the first time (this is hopeless for true macro-objects, since decoherence, due to its extreme efficiency, is always "immediately there"). Superpositions of the states of a few photons were generated, where the strength of the coupling to the surroundings, and thereby the influence of decoherence can be experimentally manipulated to a certain degree. Even though this is rather all going on in microscopic dimensions, these states are often called "Schrödinger-cat states". The lifetime of these "cats" is actually a few microseconds. For "true" cats the decoherence times would be smaller by many orders of magnitude (in particular much smaller than the lifetime of the cat), so that such objects always appear to be in one of their classical states. This means that the infamous "quantum jumps" are a consequence of extremely short decoherence times. In reality, they proceed continuously, just extremely rapidly.

The transition from quantum-mechanical behavior to classical behavior can be summarized in the following scheme.

decoherence in space	
theory	*experiment*
superposition of positions	double slit
⇓	⇓
ensemble of positions	"particles" (Wilson chamber)

decoherence in time	
theory	*experiment*
superposition of different decay times	"collapse and revival" — experiments
⇓	⇓
ensemble of different decay times	"quantum jump" — experiments

In both cases, the ability of quantum states to interfere is destroyed by the interaction with the environment and thereby a "classical" description is made possible.

The structure of space-time

Very similar considerations can be applied to gravity. According to the general theory of relativity, space, time, and gravity are very closely linked with each other (gravitational forces can be understood in geometrical terms as curvature of space and time). All experiments carried out so far support the assumption of a well-defined classical space-time structure down to distances of 10^{-15} cm. If gravity is also subjected to the laws of quantum theory (which appears inevitable for reasons of consistency), the superposition principle would allow any superposition of different gravitational fields (different space-times). Now it has to be taken into consideration that space-time is coupled to matter, which again is "measuring" the former. The mechanisms described above are also applicable here.

A simple example is shown in Fig. 8.10: the value of the gravitational acceleration g inside a cube is "measured" by molecules that pass through the cube, because their trajectory depends on the value of g (see box "decoherence and gravity").

For physicists: decoherence and gravity

Consider as a heuristic model a cube with the length of the edges L (see Fig. 8.10), and a homogeneous gravitational field that is de-

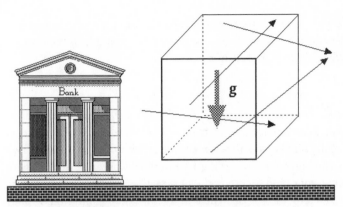

Fig. 8.10 The value of the gravitational acceleration in a certain spatial region influences the trajectory of molecules that cross this region. In this way matter is always and inevitably measuring the gravitational field.

scribed by a superposition of two accelerations g and $g\prime$

$$|\Phi\rangle = c_1 |g\rangle + c_2 |g'\rangle$$

Each particle χ that moves through this volume "measures" the value of g, because the actual trajectory depends on g:

$$|g\rangle |\chi(0)\rangle \rightarrow |g\rangle |\chi_g(t)\rangle$$

Through this mechanism the coherence between g and g' is destroyed. Due to the high number of interactions — each one inefficient by itself — the density matrix $\rho(g, g')$ for the value of g changes in accordance with

$$\rho(g, g', t) = \rho(g, g', 0) \exp[-\Gamma (g - g')^2 t]$$

with

$$\Gamma = nL^4(\pi m/2kT)^{3/2}$$

for a gas with a particle density n and temperature T. For air under normal conditions and $L = 1$ cm and $t = 1$ s, for example, a relative coherence of only $\Delta g/g = 10^{-6}$ remains. The coupling between the gravitational field and matter leads in this way to the result that space-time appears to us as being always classical.

Further considerations require a theory of quantum gravity.

A deeper analysis that goes beyond this simple example requires a full-fledged quantum theory of gravity, for example in the form of the Wheeler–de-Witt equation,

$$H|\Psi\rangle = 0$$

The wave function in this equation describes all forms of matter as well as gravity (space-time geometry). The Wheeler–de-Witt equation is formally similar to a time-independent Schrödinger equation. Two new major problems regarding its interpretation arise here: Where has time gone, and where is the observer when the entire universe is described by $|\Psi\rangle$, in particular, in cosmological applications? Apart from these interpretational problems, considerations similar to the above can be made by analyzing simple models for the interaction of quantized matter and quantized space-time. The coupling to matter leads to an extremely strong suppression of interferences in the density matrix of the metric (space-time geometry) in normal situations (that is, far from the Big Bang or outside black holes). For example, one can construct solutions for a simple model of a quantized Friedmann universe that describe a superposition of an expanding and a collapsing universe. But due to the coupling to different degrees of freedom of matter and space-time geometry, the interference term is suppressed by a factor

$$\exp\left(-\frac{\pi m H_0^2 a^3}{128}\right) \approx \exp(-10^{43}) \lll 1$$

(here, H_0 is the Hubble constant, $a \approx 1/H_0$ is the radius for the present universe, m is a typical particle mass and $\hbar = c = 1$).

8.5 What follows from all this?

What are the conclusions to be drawn from these considerations? The strong dynamical coupling of macro-objects to their environment can obviously not be ignored.[6] Due to the nonlocal features

---

6) From a historical point of view it is surprising that decoherence was overlooked for such a long time. This certainly has something to do with the fact that within the Copenhagen orthodoxy a quantum-theoretical description in the macroscopic domain was considered to be unsuitable right from the beginning.

of quantum theory, a quantum-mechanical description that is carried out in a consistent way must finally include the whole universe as a direct consequence. One might object here that a similar argument is also possible within the framework of classical physics. This analogy, however, is only partial since the new features of entangled quantum states have no parallel in a classical theory.

It should be obvious by now that macroscopic properties of *certain* objects — *namely exactly the objects that we call macroscopic* — are not properties of the objects right from the outset, as might be naively assumed, but they are *generated* only through the *irreversible* interaction with the surroundings. Ironically the origin of local classical properties lies buried in the nonlocality of (entangled) quantum states. The type of interaction defines *which* characteristic feature becomes "classical": Objects appear to be localized in *space*, because typical interactions depend on *position*.

Are all problems now solved by decoherence? Unfortunately not, because the rules of quantum theory were used so far just in a pragmatic way. Let's take a closer look.

What is an observer?

The advantage of the considerations presented so far was the fact that the results are independent of any interpretation of quantum theory and therefore they are largely agreed (among physicists). But the task of physics is to draw a conclusive and consistent picture of nature, and for that reason we have to examine critically whatever we have achieved so far. The density matrices used above, for example, are just an aid for the calculation of measurement probabilities. But what is a measurement? For this, recall the beginning of this contribution: there are two laws of motion in quantum theory, namely the deterministic Schrödinger equation and the stochastic collapse. Both are inseparably blended in the dynamics of density matrices. We say that quantum theory is empirically well confirmed. What does this actually mean? It is often pointed out that the final and decisive authority is the perception of an observer. This problem was recognized early as being essential in the discussions about quantum theory. For example, Heisenberg reports in his autobiography "Der Teil und das Ganze" on a conversation with Einstein, in which Einstein urges him:

... it may be of heuristic value to recall what one really observes. But from a principal point of view it is quite wrong to insist on founding a theory on observed quantities alone. In reality just the opposite is true. Only the theory decides about what can be observed. ... On the entire long path from a process up to our conscious perception we need to know how Nature is working in order to claim that we have observed anything at all.

Subjective perception has clearly something to do with our brain. Does quantum theory also hold there?

Decoherence in the brain

There are many models for the communication between neurons in the brain (neuronal networks). Practically all of them use classical pictures in the sense that neurons are always in a classically describable (in particular, local) state, which changes in the course of time following certain laws. As pointed out several times, quantum theory allows many nonclassical states, to which — when set in *parallel* with subjective perception — no obvious meaning can be attributed. From a quantum-theoretical point of view the superposition

$$|\Psi\rangle = |\text{neuron is firing}\rangle + |\text{neuron is not firing}\rangle$$

is, for example, a totally legal state (compare Fig. 8.2). The reasons why coherences have no relevance in such a state were only recently quantitatively examined by Max Tegmark.

It turned out again that the surroundings distinguish the two classical alternatives very rapidly and therefore destroy coherence (Fig. 8.11). Tegmark estimated typical time scales down to $t \approx 10^{-20}$ s. This means that we have no chance to make "strange" perceptions, because the above superposition is far too unstable. Obviously quantum theory can be successfully extended into the brain of the observer.

Quantum theory taken seriously

We now have reason enough to take the description of nature by quantum theory very seriously. Therefore, a few final remarks about

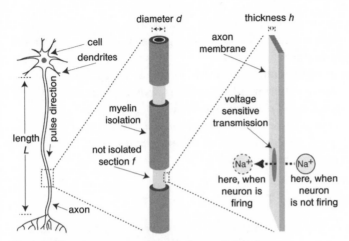

Fig. 8.11 When two neurons communicate, an electrical pulse migrates along an axon. The difference between "firing" and "not firing" consists essentially in the fact that about 10^6 sodium or potassium ions are on one side or the other of a membrane. This is distinguished very rapidly by the surroundings. (Illustration from M. Tegmark, *Phys. Rev.* **E61**, 4194.)

the question as to which interpretation of quantum theory or which alternative suggestions appear to be reasonable are in order. First, very briefly and without going into any detail, a few remarks about interpretations that from my point of view are not leading anywhere, but are frequently (often just implicitly) supported.

First, there is the widespread opinion that quantum theory is simply a statistical theory. The collapse during a measurement would then be no surprise, because it would just express the gain of information by the observer. Unfortunately, the situation is not that simple, since a wavefunction must not be considered as an ensemble of wavefunctions, because of coherence. An attempt to establish a statistical approach would require the design of a completely new theory. (Theories with additional parameters like Bohm's guiding-wave theory are not leading much further either.)

One — perhaps the preferred — method to sweep all problems under the carpet is the assertion that quantum theory is only applicable to micro-objects, while *through decree* (or should I say wishful thinking?) a classical description would be valid in the macroscopic domain. This leads to all the well-known paradoxes of quantum theory that are discussed repeatedly, but actually just because such a description is *inconsistent* (even though this is then grandly named "complementarity" or "duality" by the supporters of the Copenhagen interpretation). In addition, micro-objects and macro-objects are so strongly dynamically coupled, as we have already seen, that it is not possible to define any boundary.

When we stay consequently within a quantum-mechanical description — hence, with a description of nature by wavefunctions — only two approaches seem to remain.

First possibility: The collapse

Since we always perceive a concrete result after a quantum measurement, whereas the Schrödinger equation, as a universal law of nature, generally gives a superposition of all alternatives due to its linearity, at some moment in time all "wrong" components of the wavefunction Ψ must disappear except the one that describes the "right" perception:

$$|\Psi\rangle = \left| \begin{matrix} \text{atom} \\ \text{decayed} \end{matrix} \right\rangle \left| \begin{matrix} \text{cat} \\ \text{dead} \end{matrix} \right\rangle \left| \begin{matrix} \text{surroundings} \\ \text{see dead cat} \end{matrix} \right\rangle \left| \begin{matrix} \text{observer sees} \\ \text{dead cat} \end{matrix} \right\rangle$$

$$+ \left| \begin{matrix} \text{atom not} \\ \text{decayed} \end{matrix} \right\rangle \left| \begin{matrix} \text{cat} \\ \text{alive} \end{matrix} \right\rangle \left| \begin{matrix} \text{surroundings} \\ \text{see living cat} \end{matrix} \right\rangle \left| \begin{matrix} \text{observer sees} \\ \text{living cat} \end{matrix} \right\rangle$$

$$\xrightarrow{\text{collapse}} \left| \begin{matrix} \text{atom} \\ \text{decayed} \end{matrix} \right\rangle \left| \begin{matrix} \text{cat} \\ \text{dead} \end{matrix} \right\rangle \left| \begin{matrix} \text{surroundings} \\ \text{see dead cat} \end{matrix} \right\rangle \left| \begin{matrix} \text{observer sees} \\ \text{dead cat} \end{matrix} \right\rangle$$

This means nothing less than an explicit modification of the Schrödinger equation. Such a correction must obviously not happen too early in the microscopic domain, because otherwise it would have been experimentally observed. But it must happen before subjective perception comes into play. There have been many suggestions to accomplish such a collapse. Already von Neumann proposed a connection between collapse and the consciousness of the observer, an idea, which was later picked up by Wigner and others. Models that explicitly modify the Schrödinger equation were discussed, for example, by Penrose (who wants to hold gravity responsible for the collapse) and first of all in detail by Ghirardi, Rimini and Weber. However, there is no experimental indication so far that the Schrödinger equation would anywhere lose its validity.

Second Possibility: The Everett interpretation

When the Schrödinger equation is indeed valid without restrictions, all components of the universal wavefunction Ψ remain in existence. A generalization of the state above would describe, for example, the decay of an atom ("quantum jump") with different decay times,

$$|\Psi\rangle = \sum_t \left|\begin{matrix}\text{atom decays} \\ \text{at the time } t\end{matrix}\right\rangle \left|\begin{matrix}\text{Geiger counter} \\ \text{responds} \\ \text{at the time } t\end{matrix}\right\rangle$$

$$\left|\begin{matrix}\text{surroundings} \\ \text{see dead cat} \\ \text{at the time } t\end{matrix}\right\rangle \left|\begin{matrix}\text{observer} \\ \text{sees dead cat} \\ \text{after the time } t\end{matrix}\right\rangle$$

Since each component of this wavefunction (each term in the sum) also contains the observer, more and more different versions of the same observer are generated in the course of time, with different perceptions and different (but in itself consistent) memories about his perceptions. The structure of the world is therefore extremely rich (John Bell once called this multitude of components "extravagant"), but it follows automatically from a simple fundamental dynamics, namely the Schrödinger equation.

What is the better choice?

Which one of the two possibilities should be preferred? Both have their strengths and weaknesses, some of them are compared in the Table 8.3.

Collapse models must specify at which place the deviation from the Schrödinger equation occurs precisely. Only a few explicit suggestions exist for that so far. What triggers the collapse? Is it consciousness itself? Or is the collapse already happening in the "physical domain", is perhaps gravity — which is still causing big problems regarding the inclusion into a general quantum theory — the reason? The attractiveness of collapse models lies in the fact that subjective perception can still directly be linked to physical states (in the brain). In contrast to this, subjective perception in the Everett interpretation has to be related to certain *components* of the wavefunction. An essential conclusion is then the parallel existence of many different versions of each observer, all with different perceptions (and memories). This is an unavoidable consequence, which many people do

Table 8.3 Collapse models and Everett interpretations have similar problems. Though their predictions are different in principle, an experimental distinction appears to be extremely difficult.

Collapse models	Everett
How and when does a collapse occur?	What is the precise structure of the Everett branches?
traditional psychophysical parallelism: perception is parallel to the *state* of the observer	new form of psychophysical parallelism: perception is parallel to a *component* of the universal wavefunction
probabilities are postulated	probabilities also require additional axiom (controversial)
perhaps problems with relativity theory	no problems for local interactions
experimental test:	experimental test:
search for collapse-type deviations from the Schrödinger equation	search for superpositions in the macroscopic domain
⇓	⇓
appears to be impossible because of decoherence	appears to be impossible because of decoherence

not like to come to. But it has to be underlined that this seems to be inevitable when the Schrödinger equation is really the fundamental law of nature.

A frequently asked question is about the origin of probabilities in quantum theory. In collapse theories, these are simply inserted "by hand", which means that these theories contain some kind of randomness that does not need to be explained any further. In Everett interpretations *all* possibilities are realized in parallel. Nevertheless, it is still controversial to what extent the numerical values for probabilities, which quantum theory predicts via collapse (see Table 8.2), come out right without additional assumptions.

Some collapse theories lead to serious problems with relativity theory. This is not surprising, because a wavefunction with whatever spatial extension can be changed by a collapse just anywhere. Though this is not a concern for all theories of this type, this problem does not occur at all in the Everett interpretation, because here all interactions and with that the whole dynamics are always local.

Since physics is based on experiments one would like to have a test in order to be able to distinguish between these two actually very different approaches. This is unfortunately not all that simple, even though both theories give, in principle, different predictions. In order to test collapse theories, deviations from the Schrödinger equation should be investigated. But the effects of decoherence are just in a way that they *look* like a collapse. A distinction is therefore extremely difficult. On the other hand, the Everett interpretation claims that a collapse never occurs and therefore all components of the wave function always continue to exist. This means, among other things, that a superposition of dead and living cat is never destroyed, from a global point of view. Unfortunately, such extreme superpositions are locally unobservable — due to decoherence. Therefore, the interpretation problem of quantum theory still exists, but now it appears in a different light.

References

A very readable review article was written by Max Tegmark and John A. Wheeler, "100 Years of Quantum Mysteries" published in *Scientific American* Feb. 2001, p. 54.

The monography "Veiled Reality" from Bernard d'Espagnat (Addison-Wesley, 1995) presents a careful and detailed discussion of the interpretation problem of quantum theory.

All aspects of decoherence are examined in the book "Decoherence and the Appearance of a Classical World in Quantum Theory" by E. Joos, H. D. Zeh, C. Kiefer, D. Giulini, J. Kupsch, I. O. Stamatescu (Springer-Verlag, 2003).

Further literature and information about decoherence can also be found at *www.decoherence.de*.

9

Quantum information processing: Dream and Realization

Rainer Blatt

Computers have been available for several decades now and they are ubiquitous in all areas of our societies, and have changed our daily lives profoundly. The advances in producing semiconductors during the last 40 years have resulted in a computational power on our desks that could not have been dreamed of by earlier generations. In 1949, a journalist wrote about the then most powerful computer ENIAC in the magazine "Popular Mechanics": *"Where a calculator on the ENIAC is equipped with 18 000 vacuum tubes and weighs 30 tons, computers in the future may have only 1000 tubes and weigh only 1 ½ tons."* This futuristic view became reality all too soon and today is surpassed by orders of magnitude, computers became different and faster than foreseeable in those days. The rapid development of computers was empirically described first by Gordon Moore (one of the founders of INTEL corporation) who noticed in the 1960 s that the number of elements in a computer doubles approximately every 18 months to two years, a rule that is nowadays known as Moore's law. Quite surprisingly, this law is still valid today and provides the basic guideline for the semiconductor's roadmap, which industry needs to adhere to in order to stay competitive.

Of course, the evolution becomes possible only if the required building blocks, memory and processor units, can be made smaller and smaller. Accordingly, the necessary efforts increase and correspondingly, there is increasing interest in ever-smaller structures using nanotechnologies. While in the 1960s macroscopic amounts of matter (i.e. about 10^{17}–10^{18} atoms, approximately several 10 μg of Silicon) were used to implement the basic element of information

Entangled World, Jürgen Audretsch
Copyright © 2005 WILEY-VCH Verlag GmbH & Co. KGaA, Weinheim
ISBN: 3-527-40470-8

storage (the bit), in 2001 only about 10^7 atoms are required for one bit. Richard Feynman commented on this development in 1986 with the words "... *we are to be even more ridiculous later and consider bits written in one atom instead of the present* 10^{11} *atoms. Such nonsense is very entertaining to professors like me. I hope you will find it interesting and entertaining also.*"

If the miniaturization goes on at the same rate it can be expected that in about 10–15 years from now we will have reached the limit where a single atom will have to suffice for storing one bit. A similar estimate is obtained by calculating the number of available electrons per electronic element and it becomes obvious that sooner rather than later the laws of quantum mechanics must be employed in order to describe the storage and manipulation of information.

Inspired by such deliberations, R. Feynman [1] and D. Deutsch [2] first proposed computers explicitly using the laws of quantum mechanics. According to Feynman, some quantum-mechanical calculations could be implemented directly on a quantum computer without using complicated procedures on a classical computers. However, such ideas were considered remote since a quantum computer did not exist and was not considered realistic at that time. Worse, not a single algorithm existed that would have motivated the use of a quantum computer.

This view changed drastically in 1994 when P. Shor [3] proposed a procedure, or more precisely, a quantum algorithm that enables one to factor large numbers much faster than with any (known) classical computation. It was realized with considerable excitement that a quantum computer technology would, for example, endanger a good deal of today's powerful cryptosystems since they are all based on the fact that factorization of large numbers is a difficult computational problem. While the factorization of a number N requires efforts that grow exponentially with the number L of the digits ($\sim e^{L^{1/3}}$), the Shor algorithm needs only polynomially growing efforts ($\sim L^2$). This algorithm demonstrated for the first time the potential power of a quantum computer and spawned a worldwide search for implementing such a device. In 1997, Lov Grover found a fast algorithm [4] for searching databases. This is a procedure, which is quite generally applicable, and solves, for example, the telephone book problem: Searching the phone number of a person is an easy task if one knows the person's name. On the other hand, knowing the phone number requires in the worst case searching the entire phone book in order to find, for example, the name or the corresponding address, i.e. it requires efforts that grow as the number of entries. Grover's algorithm, instead, requires only efforts that grow as the square root of the number of entries.

Today, a number of quite advantageous algorithms and applications are known, most of them in the field of number theory. Aside from this, the original ideas of Feynman, i.e. using a quantum computer for the simulation of quantum problems become increasingly interesting today. And eventually, a quantum computer will provide a means to synthesize and design special quantum states that may be particularly suited for improved precision measurements.

9.1 Computation — a physical process

Information processing by a computer and also by a quantum computer[1] is always represented by a physical process that must be accordingly implemented by physical means, i.e. it must be realized by hardware components. Thus, let us recall the basic ingredients and the function of a classical computer prior to describing the quantum version.

First, information must be provided and stored physically, then after initializing and setting the input of the computer, the computation itself must be carried out by a sequence of precisely defined logical operations, all of which are realized by physical processes manipulating the information stored, e.g. in the register. The sequence of operations, that is the computer program, is defined by the algorithm solving the problem at hand, and is a precise recipe for the necessary physical operations. Whether this corresponds to shifting the pearls of an abacus or to controlling currents and voltages inside an electronic chip, information processing is always physical. Eventually, the result of the computation must be read and this again requires a physical measurement and the observation of the final state, i.e. the physically represented outcome of the procedure.

In classical computers, information is stored with electronic memories, representing either a logical 0 or 1. This is the well-known

1) Compare to Sections 4.5, 6.2 and 7.4.

"bit" (from "*bi*nary dig*it*") and with bit rows they form registers and then whole memory banks. Input and output are realized by setting and measuring of the corresponding register state. Computation is achieved using sequences of logical operations, steered by a program unit (cf. Fig. 9.1).

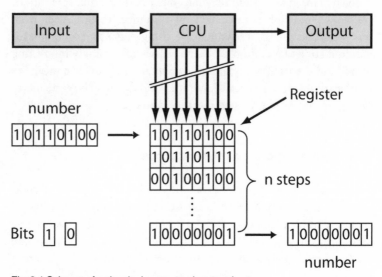

Fig. 9.1 Scheme of a classical computer: input and output numbers are encoded by a register of bits (from *bi*nary dig*it*), each bit consisting of two logical states, either 0 or 1. A central processing unit (CPU) implements an algorithm by processing the information in a register step by step using logical gate operations in a sequence defined by a program.

Generalizing this concept and using the laws of quantum mechanics consequently means that the smallest unit of information can also exist as a superposition of two classical states 0 and 1. To describe this we write the wavefunction

$$|\Psi\rangle = c_0|0\rangle + c_1|1\rangle \tag{9.1}$$

that denotes the state of a system, i. e. the information on a system as a superposition of two possible quantum states $|0\rangle$ and $|1\rangle$. A measurement of such a storage site then reveals the state $|0\rangle$ with probability $|c_0|^2$ or state $|1\rangle$ with probability $|c_1|^2$. This is the basic unit for storing quantum information and is called a *"qubit"* (short for quantum bit). Accordingly, a row of such qubits then constitutes a *quantum register*.

Corresponding to the laws of quantum mechanics, such a quantum register is fundamentally different from a classical register. While classical bits in a conventional computer are independent of each other, quantum bits forming a quantum register usually cannot be considered in that way. Rather, the quantum state of such a register must be written as the general superposition of all possible quantum states of the individual qubits. For example, for a quantum register comprised of 3 qubits one obtains the following wavefunction:

$$\begin{aligned}
|\Psi\rangle &= c_{000}|001\rangle + c_{001}|001\rangle + \cdots + c_{110}|110\rangle + c_{111}|111\rangle \\
&= c_0|0\rangle + c_1|1\rangle + \cdots + c_7|7\rangle
\end{aligned}$$

In this case, one obtains the superposition of 8 different states that are combined from the 3 qubits of the quantum register. The number of possible states of the quantum register with L qubits grows as 2^L and the state of a quantum register can no longer be written as the product of the basis states. States of this kind are known in quantum mechanics as being *"entangled"*, and the quantum register must be considered to be in an entangled state.

The fact that qubits and quantum registers need to be treated in a completely quantum-mechanical way has some dramatic consequences. In particular, superpositions of states remain unaffected only if the system under consideration is not touched nor observed in any way. For example, each attempt to observe the content of a qubit or even to track information processing in it results in an immediate projection of the system onto its eigenstates, i. e. one finds either state $|0\rangle$ or $|1\rangle$. Thus, any superposition will be destroyed during that kind of action and quantum information in its general sense ceases to exist. An entire quantum register is even more sensitive and the whole state information gets lost upon the observation of even a single qubit. Thus, quantum information processing requires the "coherent" processing of the algorithms, these are processes that are completely reversible and do not reveal any information about the coefficients c_i of the pertinent qubits. Only a final measurement would reveal the outcome of the quantum information processing. This information then is obtained by reading the state of the individual qubits that constitute a quantum register.

Realizing these constraints, the physical requirements for a quantum information processor become evident. First, a quantum memory is required with which a quantum register can be formed. For coherent information processing it is necessary that the appropriate quantum states are unperturbed by unwanted external influences or observations. This, in particular, requires isolation and controlling the interaction of the quantum system to its environment. In order to prepare the quantum register in all possible states it will be necessary to initialize the quantum register to an arbitrary (initial) state and to have the individual qubits interact with each other in a controlled way. Eventually, the system must allow for a measurement of the quantum register with sufficient detection efficiency to read the outcome of a computation reliably.

9.2 How does a quantum computer work?

Once the required hardware and input and output procedures are available, the actual computation must be implemented. In classical computers these are realized using logical gate operations controlled by a central processing unit (CPU) using bit manipulations of the register, e.g. with AND, NAND, OR and XOR gate operations. For a quantum system, it will be necessary to use coherent operations for manipulating qubits. In order to demonstrate the operation of a quantum computer in principle, it is sufficient to implement the basic building blocks that allow one to break down any algorithm into a sequence of such *quantum gate* operations. Such a set of *universal gate operatons* is attained by realizing i) single-qubit operations, and ii) two-qubit operations. Single-qubit operations allow one to prepare an individual qubit of a quantum register in an arbitrary superposition state, while for the two-qubit operation the equivalent of the classical XOR operation is required. The latter operation concerns a target qubit whose state is affected by the state of a controlling qubit. This coherent operation is known as a *controlled NOT* or CNOT-gate operation. Its truth table (cf. Fig. 9.2) is the same as for a XOR-gate operation, however, it must be valid for any quantum-mechanical superposition of states and not just any two discrete inputs. Under such an operation the state-amplitude of the target qubit is flipped if and only if the state-amplitude of the controlling qubit is nonzero, otherwise the state amplitudes remain unaltered. It was shown that one can implement arbitrary algorithms with just these two operations, i.e. this set of quantum gates is a universal set.

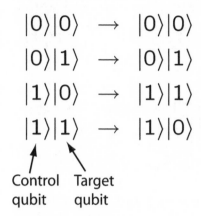

$$|0\rangle|0\rangle \;\rightarrow\; |0\rangle|0\rangle$$

$$|0\rangle|1\rangle \;\rightarrow\; |0\rangle|1\rangle$$

$$|1\rangle|0\rangle \;\rightarrow\; |1\rangle|1\rangle$$

$$|1\rangle|1\rangle \;\rightarrow\; |1\rangle|0\rangle$$

Control Target
qubit   qubit

Fig. 9.2 Truth table for a CNOT-gate operation. The state of a target qubit is changed if and only if the controlling qubit has a state amplitude $|1\rangle$. Single-qubit operations allowing for arbitrary individual qubit manipulation and the CNOT-gate operation together constitute a universal set of gate operations, i.e. arbitrary algorithms can be implemented using these operations.

Correspondingly, quantum information processing can be visualized in the following way (cf. Fig. 9.3): First, the quantum register is initialized and the input information is prepared in the individual qubits. Then a series of (coherent) single-qubit and two-qubit operations is applied to the register according to the actual algorithm to be implemented. Thus, an initial superposition is transformed into a final superposition and the entire computation is described by a unitary operation, i.e. it is completely reversible. Eventually, after quantum information processing is finished, the outcome is obtained by measuring the state of the quantum register. Of course, in that process all coherences are destroyed and the qubits are projected to their eigenstates $|0\rangle, |1\rangle$ from which the result of the computation must be inferred. Therefore, it will be the task of a programmer to include corresponding coherent operations in the gate sequence when a certain superposition is wanted or needed as the result of a quantum computation.

9.3 Which technology is suitable for a quantum processor?

Understanding the principle of quantum information processing allows us now to formulate the general requirements for a physical

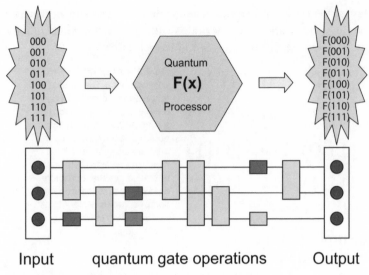

Fig. 9.3 General scheme of a quantum computer: after the initial-state preparation (input) that generates a superposition of the input states the computation is realized by a sequence of single-qubit (small boxes, different color indicates different state rotations) and two-qubit gate operations. As the result of the quantum information processing one obtains another superposition state $F(x)$. The output is obtained by measuring the final state of the quantum register.

implementation. These have been realized and noted by many authors, however, today they are known as the *DiVincenzo criteria* after David DiVincenzo of IBM who first summarized and discussed their necessity in a concise and comprehensive way [5]. In short, the storage and the processing of quantum information requires a quantum system that

- is comprised of well-defined qubits, allowing scalability.

- can be initialized to arbitrary states.

- has long-lived quantum states in order to ensure long coherence times during the computational process.

- can be coherently manipulated by a set of universal gate operations between the qubits and which is implemented using controllable interactions of the quantum systems.

- allows for an efficient measurement procedure to determine reliably the outcome of a quantum computation.

In order to scale such a quantum processor to larger devices, DiVincenzo later added two more criteria, i. e. the quantum system should

- allow for interconversion of the (memory) qubit into a moveable or so-called "flying" qubit.

- allow for a faithful transmission of the flying qubit between specified locations.

During recent years, a large variety of physical systems have been proposed and investigated for their use in quantum information processing.

Well-known quantum systems with long coherence times are, for example, given by electronic excitations of atoms and molecules. In particular, qubits may be realized with electron and nuclear spin states. Therefore, atomic systems were among the first to be proposed for quantum information processors, represented by long-lived excitations of single atoms. Already in 1995, Ignacio Cirac and Peter Zoller [6] at the University of Innsbruck suggested employing ion traps to store atomic ions as carriers of quantum information. As a different implementation, it was investigated as to how far large ensembles of nuclear moments can be used to store and manipulate qubits. As a solid-state approach quantum dots are considered, these are artificial atoms realized by nanostructured solids that exhibit atom-like energy levels. Yet a different implementation considers single atoms imbedded in carrier material, whose interaction is controlled by nanostructured electrodes. Superconducting quantum interference devices (SQUIDs) have been proposed and used as qubits. Aside from these mainstream ideas, a variety of proposals has been made and is pursued to implement qubits, e. g., the use of 2-dimensional electron gases on a liquid He surface, spectral hole burning and more.

The respective proposed systems all have very different pros and cons, only a few of them have been able to prove their suitability beyond the theoretical concept. While for atomic systems (such as spins of atoms in traps) coherence times are quite favorable, their scalability is at best difficult and in the case of nuclear spin ensembles provably impossible. On the other hand, scalability seems quite advantageous with solid-state systems, whereas decoherence presents limits that are are hard to overcome. Moreover, in these systems measuring a single spin with a high efficiency is still an unsolved problem. In spite of these shortcomings, with NMR-based ensembles of nuclear spins a variety of quantum processes, such as a CNOT-operation or even a Shor algorithm (factorizing the number 15) have been demonstrated. Unfortunately, this technology lacks the scalability potential beyond, say 10–15 qubits, and therefore this technology is not considered a viable implementation for a quantum processor.

While all of the technologies considered above have their short-comings, it turned out during the last few years that currently the trapped-ion technology seems quite favorable in that the technology fulfills all DiVincenzo criteria. Moreover, several key experiments were able to clearly demonstrate already quantum information processing with a few qubits and scalability, although technically difficult, seems possible with existing technology.

9.4 Quantum computer with trapped ions

Of all the currently investigated approaches, the use of strings of trapped ions in electromagnetic traps, so-called Paul traps, seems currently to provide a viable route towards quantum information processing and there are several investigations underway for implementation of a rudimentary quantum computer based on this technique. In the following, the technique will be briefly described and the state-of-the-art will be highlighted with the latest experimental results.

Experiments with single atoms became possible at the beginning of the 1980s, after early experiments had demonstrated that single ions could be stored in a certain configuration of electromagnetic fields. With these, single atomic ions became available for repeated preparation and (quantum-mechanical) measurement. For the confinement of these ions inhomogeneous electrical radio-frequency (rf) fields are used that are generated with specifically shaped electrodes forming the actual trap. As is indicated in Fig. 9.4, single ions are commonly trapped using an arrangement consisting of a so-called ring electrode and two endcap electrodes. With an rf-field applied to these surfaces a single charged atom will move towards the correspondingly (oppo-

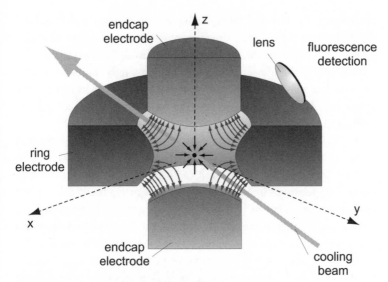

Fig. 9.4 Paul trap for the confinement of single ions. The trapping rf-potential is applied between the (connected) endcap electrodes and the ring electrode. With appropriate voltages and frequency a single particle experiences on average a force towards the center of the trap. Excitation of resonance for optical detection is achieved using a laser beam that is also employed for optical cooling of the confined particle.

sitely) charged electrode. Choosing appropriate voltage and frequency of the rf-field the motion of the ion will be such that it never actually reaches one of the electrodes, it rather tumbles back and forth between the electrodes and so stays within the trap volume formed by the metallic surfaces. Thus, a single ion remains available in free space for extended and repeated measurements. Such trapping configurations have been known since 1958 and are named after their inventor, Wolfgang Paul (University of Bonn, Nobel prize 1989), as Paul traps and are frequently used in mass spectrometry. However, only since the early 1980s it became possible to store and optically detect single ions in Paul traps. This is due to the fact that the confined ions usually move quickly inside the trap, so that they exhibit a large Doppler broadening and accordingly are difficult to detect spectroscopically. With the advent of the optical cooling techniques, originally proposed by Hänsch and Schawlow for free atoms and by Wineland and Dehmelt for trapped particles, then first realized by

Neuhauser, Toschek and Dehmelt (Univ. Heidelberg) and Wineland (NBS Boulder, Co) with a single trapped ion, the motion of a single ion could be reduced such that its amplitude is smaller than the wavelength of the incident radiation. Thus, a single trapped ion scatters continuously many photons that can even be observed with the naked eye. These experiments laid the foundation of experimenting with single atoms and their manipulation using laser light.

Due to the fact that such confined particles in ultrahigh vacuum do not collide with walls or a surrounding gas, it became possible to observe very narrow (optical and radio-frequency) transitions with single trapped ions in Paul traps. Consequently, these devices are nowadays routinely used for precision experiments, in particular for atomic clocks, where the uninterrupted and unperturbed temporal evolution of superposition states serves as a "precision pendulum" for the realization of the SI second. In view of the general requirements for a quantum computer it is obvious that the ion-trapping technique seems well suited for an implementation of quantum information processing.

One of the unique features of single-atom experiments is that they are available over and over again for repeated measurements. In the middle of the 1980s single ions in Paul traps for the first time allowed the observation of the long-postulated phenomenon of quantum jumps. In such an experiment, one exploits the fact that during the scattering of fluorescent photons the involved electron must jump back and forth between the corresponding electronic states. However, if during this process the electron gets excited to a different (third) electronic state it is no longer available to generate fluorescence and thus the atoms ceases to radiate. Thus, the resonance fluorescence exhibits characteristic jumps (cf. Fig. 9.5) indicating the projection of the electron to and from the third atomic state. If the atom scatters light, then prior to its observation the atom was in a state accessible to the scattering (monitoring) laser. On the other hand, if no fluorescence is observed, then the atom was excited and the electron is available to shuttle back and forth between the fluorescing states. In this way, a single absorbed photon (exciting the electron to a nonradiating state) causes many photons not to be scattered on the monitoring transition and this allows one to detect the quantum state of a single atom with nearly 100 % efficiency. This technique, known as *electron shelving* or the *quantum jump* method results from experimenting with a single particle and provides an important asset for realizing a quantum information processor.

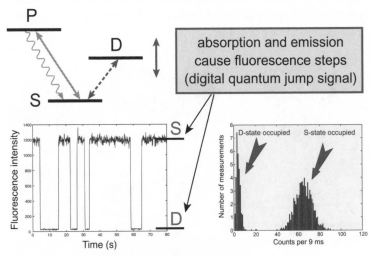

Fig. 9.5 Spectroscopy with quantized fluorescence: continuous excitation on a strong transition excites resonance fluorescence that is optically detected while the electron shuttles between states S and P. If the electron gets excited to the D level, it is no longer available on the S–P transition and fluorescence ceases. Once the electron "jumps" back to the S state, fluorescence starts again. Counting the fluorescence photons for a certain time while probing the atom on the S–P transition then reveals the histogram shown to the right. Thus, the internal state of the atom (S or D) can be observed with nearly 100 % detection efficiency.

Qubits are thus implemented with a two-level system encoded in a single trapped ion, quantum registers need rows of ions that are interacting in a controlled way. This can be realized using so-called linear Paul traps (cf. Fig. 9.6). Such a trap device is derived from the Paul mass filter that employs four hyperbolic electrodes and rf-voltages to transmit an ion beam in a mass-selective way. For a trap, one uses an arrangement of four rods (rf-voltages) and two ring electrodes (dc-voltages) for axial confinement. With such a device ions can be confined along the longitudinal axis in the center of the 4 rods and under the action of laser cooling, the ions form a string as shown in Fig. 9.6. The distance between neighboring ions, however, is slightly varying since each ion sees a different repelling charge of its neighbors. Thus, an ion string provides a register with the quantum information stored in the individual electronic states of the ions. The distance between neighboring ions can be adjusted such that they can be manipulated individually with a focused laser beam, i. e. single-qubit operations are possible.

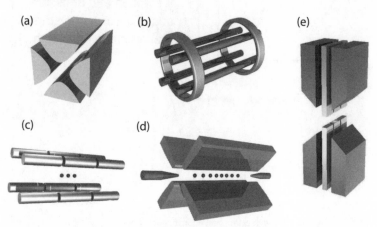

Fig. 9.6 Linear Paul traps are all derived from the Paul mass filter potential (a) that allows one to transmit an ion beam in a mass-selective way through the longitudinal axis using appropriate rf voltages and frequencies. With additional (ring) electrodes (b) ions can be confined along the trap axis. The rings can be replaced by segments (c) or tips (d) and even combinations of retracted electrodes and segments (e) can be used.

In this way, already three of the DiVincenzo criteria are fulfilled. In 1995, Ignacio Cirac and Peter Zoller (Univ. Innsbruck) proposed a procedure to implement the required CNOT gate operation in order to realize a universal set of quantum gates. The important idea here was that the ions interact with each other via the repulsive Coulomb forces and thus the ion motion provides an additional degree of freedom that can be used to carry and convey information. Accordingly, the Cirac–Zoller proposal is, in short, to use the Coulomb interaction to achieve a state change of a target qubit conditional to the state of a controlling qubit. For this purpose, they suggest that one starts with the quantum register at complete rest, i. e. initially the harmonic motion of the string of ions needs to be completely frozen, or in quantum-mechanical terms, it is considered to be in the ground state of the corresponding harmonic oscillator. Thus, the initial state vector of the quantum register is considered to be

$$|\Psi\rangle = \sum_{\underline{x}} c_{\underline{x}} |x_{L-1}, \ldots, x_0\rangle \otimes |0\rangle_{CM} \qquad (9.2)$$

where the vector $\underline{x}$ denotes all internal electronic states of the qubits and the harmonic oscillator (center of mass) state is exactly $|n\rangle_{CM} = |0\rangle_{CM}$, i.e. the ground state. With a laser pulse directed to the controlling qubit, the internal excited state information of that particular qubit (which could be any ion of the quantum register, depending on the algorithm in question) is then mapped onto the the center-of-mass motion, i.e. when there was a (nonzero) excited-state amplitude, that is now written into an excited-state amplitude of the motion, $|1\rangle_{CM}$ whereas the controlling ion's internal state is put in its ground state. Due to the Coulomb interaction then this quantum of center-of-mass motion describes the movement of the entire ion string and is, of course, shared by all ions. Thus, the new quantum state of the register is an entangled state of the internal (electronic) and the external (vibronic or motional) states. With a laser directed then to any other ion (serving as the target ion, depending on the algorithm) it is then possible to manipulate the internal state of that target ion without changing its motional state if and only if there is motion in the ion string. Finally, undoing the first step to the controlling ion, i.e. taking the motion out of the string and restoring the controlling qubit's amplitude makes it possible to realize the truth table of the CNOT-gate operation completely and coherently for all amplitudes. Together with arbitrarily possible individual qubit manipulations the implementation of such a CNOT-gate operation constitutes a universal set of gate operations that allows one to realize any quantum computation. Furthermore, and as shown by Cirac and Zoller already in their seminal paper in 1995, this approach is actually scalable, i.e. adding more and more qubits does not add exponential difficulties, at least not in principle.

As a carrier for the quantum information, long-lived atomic states are required. Since the proposal of 1995 a variety of ions has been investigated as candidates for quantum information processing. As indicated in Fig. 9.7, the qubits can either be encoded using narrow optical transitions, e.g. given by so-called forbidden transitions, or by using radio-frequency transitions between Zeeman-split levels or levels split by hyperfine interaction, similar to those used in

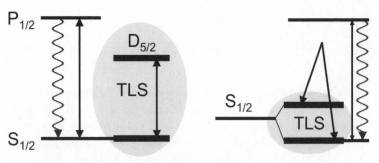

Fig. 9.7 Two-level systems (TLS) available with trapped ions. Left: optical transitions on forbidden lines (quadrupole transitions, intercombination lines), e.g. S–D transitions in earth alkalis such as Ca^+, Sr^+, Ba^+, Yb^+, Hg^+, etc. using single-photon excitations. Right: microwave transitions (hyperfine and Zeeman transitions) in odd earth alkalis, e.g. $^9Be^+$, $^{25}Mg^+$, $^{43}Ca^+$, $^{87}Sr^+$, $^{111}Cd^+$, $^{137}Ba^+$, $^{171}Yb^+$ driven by Raman transitions.

atomic clocks. While the actual technical implementation certainly varies strongly depending on the ions used, the Cirac–Zoller concept can be realized either way. During the last few years, in particular the experiments of the Boulder group (D. Wineland et al. at NIST, Boulder, Co., USA) and of the Innsbruck group (R. Blatt at Univ. Innsbruck, Austria) have proved the feasibility of quantum information processing using different approaches. While the Innsbruck experiments follow closely the original Cirac–Zoller proposal and use optical transitions with individual addressing in Ca^+ ions, the Boulder experiments use Be^+ ions where the qubits are encoded in hyperfine states and coherent manipulation is achieved using Raman transitions with limited addressability. Although quite different experimentally, both experiments have implemented quantum information processing quite successfully and are very similar conceptually. Below, we will discuss and present some examples based on the Innsbruck Ca^+ experiment, however, the Boulder results are very similar and at this time it is by no means obvious which element will actually be best suited for a scaled implementation using trapped ions. Figure 9.8 shows the Ca^+ setup schematically and a string of ions representing a quantum register.

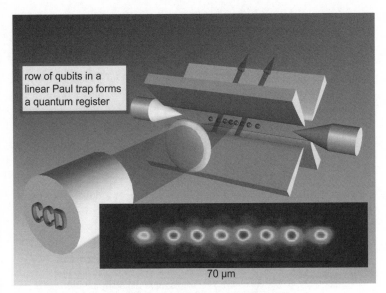

Fig. 9.8 Ca$^+$ ions are stored in a linear Paul trap. The inset shows a quantum register with eight ions that can be individually addressed and manipulated. Resonance fluorescence is recorded with an imaging CCD camera.

9.5 Quantum information processing with a single ion

The primitives of quantum information processing are given by single-qubit rotations. By this, we mean the manipulation of a single trapped ion using a laser pulse. In order to understand these operations more clearly, let us consider first that the single two-level system is residing in a harmonic trap. Thus, the quantum-mechanical description of the joint system needs to take this into account. Figure 9.9 sketches this situation: whereas the atom is described by the two levels $|g\rangle$ (ground state) and $|e\rangle$ (excited state) only, the quantized harmonic motion provides infinitely many levels $|n\rangle$ of excitation, all of which are equidistant and separated by the trap frequency, i.e. the sloshing motion of the ions in the well. This system cannot be considered independent, rather each two-level system can be in any of the trap states and the entire system could be in any superposition state of all of these. In order to keep things as simple and controllable as possible, we consider the lowest states $|n, g\rangle = |n\rangle|g\rangle$ and $|n, e\rangle = |n\rangle|e\rangle$ only, i.e. we optically cool the system such that it resides in $|n\rangle = |0\rangle$. Then only three excitations with resonant laser pulses become possible. These are (i) excitations on the carrier, i.e. a laser pulse with a frequency matching exactly the energy difference between $|n, g\rangle$ and $|n, e\rangle$ and (ii) excitations on the sidebands, i.e. laser pulses with frequencies matching exactly the energy differences between $|n, g\rangle$ and $|n + 1, e\rangle$ (upper or blue sidebands) or between $|n, g\rangle$ and $|n - 1, e\rangle$ (lower or red sidebands) become possible. Thus, with carrier transitions we are able to manipulate the internal degree of freedom, i.e. the single-qubit operation required. With sideband transitions, internal and external degrees of freedom are involved, i.e. a quantum of motion is either added to (upper sideband) or taken out of (lower sideband) the system, while simultaneously the internal degree of freedom is affected.

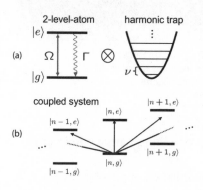

Fig. 9.9 Two-level atom confined in a harmonic trap that is represented by equidistant energy levels, different by the trap frequency ν (a). The joint quantum system consists of many levels, which are labelled as $|n, x\rangle = |n\rangle|x\rangle$ by the internal levels $|x\rangle = |g\rangle, |e\rangle$ and the trap levels $|n\rangle$. Excitations that do not change the motional state $|n\rangle$ are called carrier transitions, other transitions are named sideband transitions since they occur for a different frequency setting of the manipulating laser (b).

These three operations constitute the basic primitive pulses for quantum information processing and the computational subspace for a single ion is restricted to the states $|0, g\rangle, |0, e\rangle, |1, g\rangle$ and $|1, e\rangle$ (cf. Fig. 9.10). With the sideband pulses it will be thus possible to write internal amplitudes to the motion and back, a feature realized by Cirac and Zoller and the basis of the ion-trap quantum computer. While details of the technical implementation are beyond the scope of this book, it should be noted that special care must be taken that the sideband excitations do not leave the computational subspace.

The simplest quantum algorithms can already be demonstrated using two qubits only. Thus, even with a single ion such an experiment

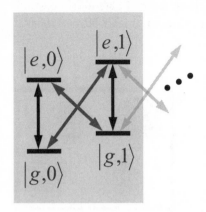

Fig. 9.10 Computational sub-space for a single trapped ion. After laser cooling the ion resides in the ground state of the harmonic oscillator $|n\rangle = |0\rangle$. Care must be taken that sideband excitations do not populate any states outside the computational subspace. This is usually achieved by using so-called composite pulses, i.e. a single laser excitation on a sideband is split into several pulses, a technique known from NMR state manipulation.

can be carried out due to the fact that the internal two-level-system (TLS) carries one qubit and the external (center-of-mass) motion is able to carry another qubit.

Consider the following, quite classical, problem: Throwing coins usually is a fair game since on the average the outcome has a 50 % probability of showing either side. On the other hand, if someone tries to cheat and uses a (fake) coin with two equal sides, such bets become extremely unfair. Therefore, prior to betting a measurement is in order to ensure fair play. However, answering the question whether a coin is fair (head on one side, tail on the other) or fake (heads or tails on both sides) requires two examinations, i.e. a look on each side. In a quantum world this is not the case due to the fact that superpositions are completely acceptable. Therefore, if a coin were represented in a quantum way (say a "quantum coin"), the information whether both sides are equal or different is clearly available in the quantum description. Accordingly, an appropriately taken quantum measurement on such a quantum coin would certainly be able to tell the difference since it could check for the superposition in question. Therefore, using a quantum processor and an appropriately written algorithm allows one to obtain the required information (fair or fake) in a single step. The associated quantum program is known as the Deutsch–Jozsa algorithm and it requires for a coin (i.e. a two-valued function) only two qubits that can be provided with a single ion. Clearly, an implementation of such a rudimentary algorithm does not solve an advanced problem, however, it is able to demonstrate quantum information processing and its advantages beyond classical computing.

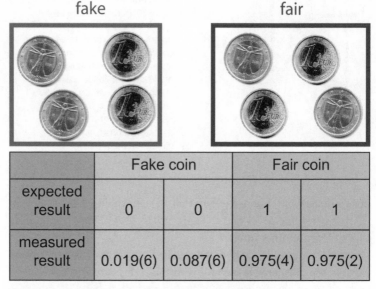

	Fake coin		Fair coin	
expected result	0	0	1	1
measured result	0.019(6)	0.087(6)	0.975(4)	0.975(2)

Fig. 9.11 Fake and fair coins can be distinguished classically only by taking two measurements, i. e. by viewing front and back. In quantum mechanics, the two sides or more precisely the information of being in either of two states can be superimposed. Thus a quantum algorithm (Deutsch–Jozsa algorithm) can reveal the information in a single computational step.

With single trapped ions quantum information processing was demonstrated using the internal and external (motional) excitations as qubits. The entire procedure requires several manipulations that are realized with a sequence of laser pulses. The outcome of the procedure is either "0" (encoded in the excited-state population of the ion) if the "quantum coin" was false (i. e. two equal sides) or it is "1" if the quantum coin was fair (i. e. two different sides of the coin). Such a measurement using a single trapped ion is shown in Fig. 9.11. Clearly, the coin information needed to be encoded into a superposition of the available states and the program sequence was adapted for that encoding, and the result is only available with an uncertainty. However, the procedure was completely general and proves that quantum information processing is faster, i.e. it requires fewer steps than classical computing. The reason for this is, of course, that in quantum mechanics a superposition is available whereas classically only two discrete states are available.

9.6 The Cirac–Zoller CNOT gate operation

As outlined above, realizing a universal quantum information processor requires the implementation of a CNOT-gate operation, i.e. a conditional operation between any two qubits of a quantum register. Following closely the proposal by Cirac and Zoller, two ions were loaded where the first one serves as the control qubit and the second one represents the target qubit. With a laser pulse directed towards ion 1, its excited-state amplitude was written to the motion using a sideband pulse, i.e. a laser pulse exciting the internal and the external degrees of freedom of the controlling qubit. Then a series of pulses was applied to the target ion in such a way that the amplitude of the wave function is changed if and only if there was any motion in the two-ion string. Finally, a laser pulse applied to the controlling ion again remaps the motional state to the excited state of the first ion and the string is at rest as it was before the entire operation started. Figure 9.12 shows schematically the sequence of laser pulses and the realized truth table of the operation.

While Fig. 9.12 shows the scheme and data from the Innsbruck experiment on Ca^+ ions, quite similar results have been obtained with a different approach and a different pulse scheme in the Boulder experiment with Be^+ ions. This clearly demonstrates that the ion-trap technique is a versatile tool to realize the basic gate operations for quantum information processing. The realized gates are universal and, as we shall see, the ion-trap scheme is scalable. Therefore, this demonstrates, for the first time, that the dream of having a quantum computer may not be so far-fetched.

9.7 Entanglement — Bell states of ions

While the use of superposition states already demonstrates some of the computational power of quantum information processing, for

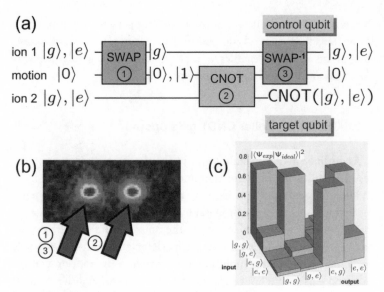

(a)

control qubit

ion 1 $|g\rangle, |e\rangle$ — SWAP ① — $|g\rangle$ ——————— SWAP⁻¹ ③ — $|g\rangle, |e\rangle$

motion $|0\rangle$ — $|0\rangle, |1\rangle$ — CNOT ② — $|0\rangle$

ion 2 $|g\rangle, |e\rangle$ ——————— CNOT$(|g\rangle, |e\rangle)$

target qubit

(b)

①③ ②

(c)

$|\langle \Psi_{exp}|\Psi_{ideal}\rangle|^2$

0.8
0.6
0.4
0.2
0

input
$|g, g\rangle$
$|g, e\rangle$
$|e, g\rangle$
$|e, e\rangle$

output
$|g, g\rangle$ $|g, e\rangle$ $|e, g\rangle$ $|e, e\rangle$

Fig. 9.12 Scheme of the Cirac–Zoller CNOT-gate operation (a). The controlling ion is addressed with a laser pulse
(b, 1) and the ion motion is excited if it carries an excited-
state population. Then the target ion is addressed (b, 2) and
a CNOT sequence is applied. Finally, the excited state of the
controlling ion is restored (b, 3) and the motion is taken out
again. The truthtable (c) clearly shows that whenever the con-
trolling ion (i. e. the first one) is in state $|g\rangle$ nothing happens
to the target ion (the second ion), whereas whenever the
first ion is in state $|e\rangle$ the second ion's state is switched. The
height of the bars indicate the "fidelity" of the operation, i. e. a
measure of how reliably the operation can be performed.

example with the Deutsch–Jozsa algorithm, its ultimate potential
becomes visible only by including nonlocal operations. Superpositions of quantum systems at different positions are outside the realm
of classical experience and therefore such 'entangled' states really
demonstrate the quantum nature of information processing systems.

Such nonlocal operations can be realized in a computational way by
making use of the Cirac–Zoller CNOT-gate operation.[2] Consider, for
example, the result of such a computation step if the controlling ion
carries a superposition as an input, clearly an operation that cannot
be done classically. As can be immediately inferred from the truth

2) For the creation of entanglement see also Section 6.3.

table (cf. Fig. 9.2), we obtain (omitting factors of $1/\sqrt{2}$)

$$|g + e\rangle|g\rangle = |(g + e)e\rangle \longrightarrow |gg + ee\rangle \qquad (9.3)$$

i. e. we obtain a highly correlated quantum state, a so-called Bell state. And using different phases for the optical pulses allows us to produce in a systematic way all Bell states given by the superpositions

$$\Psi^\pm = \frac{1}{\sqrt{2}}(|g\rangle|e\rangle \pm |e\rangle|g\rangle) = \frac{1}{\sqrt{2}}(|ge\rangle \pm |eg\rangle) \qquad (9.4)$$

$$\Phi^\pm = \frac{1}{\sqrt{2}}(|g\rangle|g\rangle) \pm |e\rangle|e\rangle = \frac{1}{\sqrt{2}}(|gg\rangle \pm |ee\rangle) \qquad (9.5)$$

Producing such states makes it impossible to describe the state in terms of individual independent particles at different locations. In fact, from a mathematical viewpoint it becomes impossible to consider the particles as independent: they are 'inextricably interwoven' and the quantum state of the system is highly nonlocal, in that it cannot be described as a sum or product of its parts. Thus, measuring one particle immediately causes a collapse of the joint wavefunction and the particles reveal a one-to-one (quantum) correlation of their respective state amplitudes. For example, using state $\Phi^\pm = \frac{1}{\sqrt{2}}(|gg\rangle \pm |ee\rangle)$ and observing, say state $|g\rangle$ on the left particle immediately implies that the right particle is in state $|g\rangle$. Such states are also called EPR states (after Einstein, Podolsky and Rosen, who first envisioned such nonlocal quantum states in 1935) and are nowadays usually called Bell states (after John Bell who used them to distinguish between the quantum and classical world). Such states are of a peculiar nature and they cannot be produced by any classical means: they require nonlocal quantum operations.

With an ion-trap quantum information processor, however, such Bell states can simply be created at the push of a button. Moreover, these states are then available as a resource for further quantum information processing, unlike the correlated pairs of photons or decaying particles, where such states are only probabilistically available and are usually destroyed when detected.

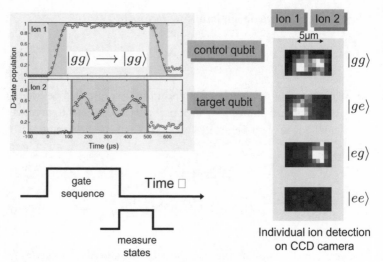

Fig. 9.13 Tracing the internal dynamics of the Cirac–Zoller CNOT gate operation with both qubits being initially in their electronic ground states. The data points show the excited state population of the qubits and the solid curves show quantum-mechanical calculations without any fitting, just taking into account the actual parameters of the experiment. The population is derived from CCD images detecting fluorescence light from single ions, as shown on the right-hand side.

9.8 Debugging the quantum processor

With a program generating Bell states, quantum information processing is already an operational reality and now is the time to think for a minute about the processing dynamics and its control. With classical computers we are usually able to follow the contents of the registers involved at any time and we have complete control over the entire process. However, with superpositions carrying the information and entanglement at work any observation of the actual content of a quantum register would immediately project the system onto its eigenstates and the quantum information held in a superposition state would be lost. On the other hand, for each given system and evolution we could interrupt the computational process at any time and then measure the momentary state of the qubits. Tracing the operation of, for example, a quantum gate allows one to compare the actual dynamical behavior of the process with a theoretical description. Thus the processes can be checked in time and thus optimized prior to routine operation of an algorithm consisting of a whole sequence of such gates. Such measurements, of course, would reveal the populations of the individual ions only, but in order to obtain the full quantum information of a given register one would need the "coherences", i.e. the coupling between the different states as well. Using single-qubit operations in addition, however, all superpositions can be rewritten into populations prior to making the measurement. Determining all entries of a density matrix

$$\rho = |\Psi\rangle\langle\Psi| \tag{9.6}$$

describing a quantum state $|\Psi\rangle$ is known as "quantum tomography", named after the well-known imaging techniques that make use of many projections to reconstruct an image. With quantum states $|\Psi\rangle$ encoded in a quantum register, prior to a (projective) measurement operations on individual ions are made by laser pulses applied locally

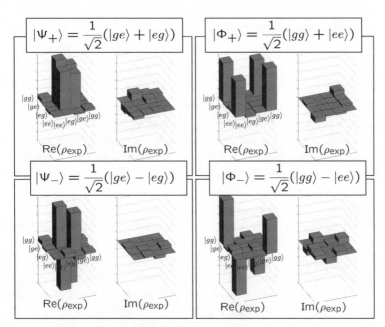

Fig. 9.14 Quantum tomography of Bell states with two trapped ions. The bars show the measured real and imaginary values of the respective density matrix elements. These Bell states were "calculated" with a fidelity of $\approx 91\%$.

to single ions and thus the elements of the density matrix ρ can be determined.

As an example of such a quantum tomography measurement, Fig. 9.14 shows the measured density matrices for the Bell states as described in Eq. (9.4). Quantum states can thus be characterized and analyzed by comparing the observed density matrix entries with ideal ones.

9.9 Processing with 3 qubits: GHZ and W states

Adding another ion to the quantum register allows us to process quantum information with three qubits. Aside from the larger computational space this makes it possible for the first time to investigate entanglement for three particles at the push of a button. Tri-partite entangled states have become important for experiments exploring the fundamentals of quantum mechanics, for quantum-communication purposes and they provide the first step towards studying multiparticle entanglement. The maximally entangled state

$$|\Psi_{\text{GHZ}}\rangle = \frac{1}{\sqrt{2}}(|ggg\rangle + |eee\rangle) \tag{9.7}$$

is known as the Greenberger–Horne–Zeilinger (GHZ) state and its importance is due to the fact that entangling more than two particles leads to a conflict with local realism for nonstatistical predictions of quantum mechanics [7] while experiments with two entangled particles and testing Bell's inequalities observe conflicts only with statistical predictions.

With the three-qubit quantum processor it is now possible for the first time to actually "calculate" these states at the push of a button and to investigate, for example, their decoherence and their dynamical behavior under the influence of a measurement. Figure 9.15 shows the quantum program to compute a GHZ state on the quantum register, the whole sequence takes about 750 µs after which GHZ states are created with $\approx 75\%$ fidelity.

Another important tri-partite entangled state is

$$|\Psi\rangle = \frac{1}{\sqrt{3}}(|gee\rangle + |ege\rangle + |eeg\rangle) \tag{9.8}$$

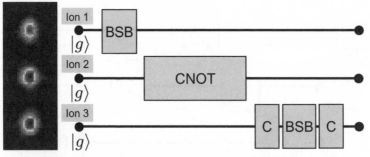

Fig. 9.15 Quantum register with three ions and algorithm to create GHZ states. The leading blue sideband pulse (BSB) on ion 1 writes a motional amplitude for the subsequent CNOT-gate operation on ion 2. For clarity, the motional qubit is no longer shown. The trailing operations on ion 3 are carrier (C) pulses and one blue sideband pulse to remove the motion from the system.

the so-called W state. It was only recently realized that states of this type represent a completely different class of entangled states. As was shown by W. Dür and I. Cirac, W states cannot be transformed into GHZ states (and vice versa) using local operations and classical communications (LOCC); they need nonlocal operations. As it turns out, GHZ and W states behave quite differently when measured and therefore quantum information processing with three qubits is already an interesting problem. Figure 9.16 shows the sequence of operations needed to create a W state within $\approx$ 400 μs with a fidelity of about 85 % every time the program is run on the quantum processor.

In this way it is possible to create a "superatom" comprised of three entangled ions with a size of about 10 μm. Due to the nonlocal character of these states they can be used to increase the sensitivity in precision measurements and for decoherence studies. The striking difference between GHZ and W states was very recently demonstrated [8] by creating these states and then investigating the result of a measurement on the central ion of the three-qubit string. Such measurements were obtained using the tomography technique and gave complete control on the information processing. While for a GHZ state the coherence, i. e. the superposition part of the quantum state was immediately destroyed upon measuring the state of the center qubit, the same measurement on a W state resulted in the creation of

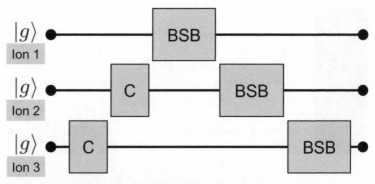

Fig. 9.16 Algorithm to create W states. The leading carrier pulses on ions 3 and 2 are followed by three blue sideband pulses (BSB) on ions 1, 2 and 3 with different pulse length and different optical phases. The sequence in this optimized form does not need a CNOT-gate operation.

a Bell state of the left and right ion. Thus, with a W state even after a partial measurement some entanglement remains, which is not the case for a GHZ case. This demonstration shows clearly how flexible these highly nonclassical states can be generated and analyzed by a computational process in a programmable way.

9.10 Conveying quantum information — teleportation

Aside from "computing states", with the availability of even a small quantum information processor, a variety of quantum protocols can now be run and tested. One of the most important and most striking ways to convey quantum information is the teleportation protocol of Bennett et al. [9]. Teleportation is concerned with the complete transfer of information from one particle to another.[3] Specifying a quantum state completely generally requires an infinite amount of information, even for qubits. Moreover, measuring a system would immediately alter its state, therefore transferring quantum information is hard. However, as shown by Bennett et al. [9], entanglement can be used together with classical communication to achieve the complete transfer of quantum information, a process coined teleportation. Teleportation using pairs of photons has been demonstrated,

3) Compare to Sections 6.2 and 7.2.

however, such techniques are probabilistic and they require postselection of measured photons.

The teleportation protocol works as follows (cf. Fig. 9.17): In order to transfer the quantum information Alice and Bob establish two channels, a classical channel and a quantum channel. On the latter they share a Bell state as a resource required for the protocol. Sender Alice rotates the unknown input two-qubit state Ψ into the Bell basis (i.e. a basis given by Eq. (9.4)) that can be achieved by a CNOT-gate operation and a single-qubit rotation. Then Alice measures in that Bell basis and obtains as a result the information that one of the Bell states $\Psi_\pm$, $\Phi_\pm$ was obtained. Clearly, during this measurement process the quantum information at Alice's location and thus the original state Ψ is destroyed. The outcome of Alice's measurement is then classically communicated to Bob who subsequently rotates his part of the originally shared Bell state depending on the measured result.

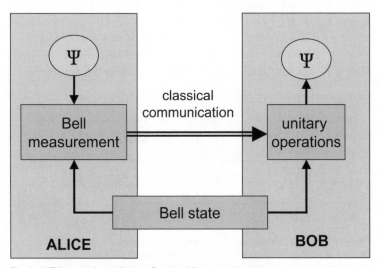

Fig. 9.17 Teleportation scheme: Sender Alice wants to transfer the quantum state Ψ to receiver Bob. They initially share a Bell state, Alice rotates her unknown input state into the Bell basis, makes a Bell measurement and communicates the outcome of this measurement via classical channels to Bob. Bob rotates his part of the Bell state according to the communicated result of the Bell measurement and thus retrieves the unknown input state.

Thus, Bob rotates his part of the shared Bell state such that he exactly retrieves the initially given state Ψ and teleportation is achieved.

With a three-qubit quantum processor, teleportation can be programmed and it generates the teleported state at the push of a button. Figure 9.18 indicates the program used for quantum information processing to run the teleportation protocol. Alice consists of two qubits, one to carry the unknown quantum information and one to share an entangled pair of states with Bob who just has the other part of the entangled pair. After initializing the qubits to their electronic and motional ground state, the Bell state shared between Alice and Bob is created. Entanglement is here used as a resource. Alice then prepares her unknown quantum state Ψ and then makes a CNOT-gate operation where Alice's part of the entangled pair (ion 2) controls the state of the qubit line carrying the unknown information (ion 1). After a single-qubit rotation (carrier operation) on ion 2 that prepares Alice's qubits for the required Bell measurement, qubits 1 and 2 are measured and the result is sent via classical communication lines to Bob. Bob, in turn, makes either nothing, an X, a Z or both operations, depending on the communicated result. X and Z denote single-qubit rotations that are given by the appropriate optical phases and pulse lengths. After these operations are completed the unknown input state Ψ is now available on qubit three while qubit 1 is undefined.

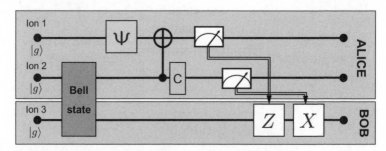

Fig. 9.18 Teleportation program: Alice has two qubits, Bob has one qubit. After sharing a Bell state, Alice prepares her unknown quantum state, rotates it into a Bell basis and makes a Bell measurement. The result of this measurement is then sent to Bob who makes appropriate rotations depending on the communicated outcome and the unknown state Ψ is retrieved on qubit 3.

Using this program, teleportation has been demonstrated with a three-qubit quantum processor using Ca$^+$ ions [10]. Similarly, albeit with a technique that uses moving ions in and out of the interaction region, the Boulder group was also able to demonstrate teleportation with three Be^+ ions [11]. In both experiments the fidelities are similiarly about 75 % and the result is obtained in the programmed deterministic way.

9.11 Towards a bigger machine — scaling the ion trap-quantum computer

One of the major advantages of the ion-trap quantum computer is its scalability. This is achieved, for example, by simply adding another ion to the string confined in the trap. Loading a third ion to the register does not change the basic frequency of the vibrating string, the lowest frequency of its common motion is always the same. Thus, by adding ions to the register the Cirac–Zoller approach works in exactly the same way as with two ions. The drawback, however, is that by adding ions the entire register becomes "heavier", and therefore influencing the ion motion becomes slower, and of course, putting everything into a ground state to begin with, becomes more and more difficult. But, more importantly, all excitations on the sideband become slower since they include the excitation or de-excitation of a vibrational quantum. Eventually, with very many ions such a system becomes sluggish and hard to handle, it will become technically very difficult to control. Therefore, this may actually limit direct scaling to tens of ions in a single string.

Consequently, it would be much better to keep single ions stored in individual sites and then try to interconnect between them. In this way the ions are well confined, isolated from each other and can be individually controlled and accessed. Quantum information processing then, however, requires a quantum channel to convey the quantum information between the different sites. One way to achieve this, is to write the static information contained in the atomic qubits to a photon that then could be transmitted via an optical fiber and finally coupled again to another ion. For this, a full transmission protocol (photonic channel) was proposed by Cirac, Zoller, Kimble and Mabuchi and a corresponding experiment is currently underway in Innsbruck (cf. Fig. 9.19.) For this, the single ions are trapped inside an arrangement of two opposing mirrors called an optical cavity that forces the atom to emit or absorb into the cavity axis. With appropriate reflection and transmission properties of the mirrors and an optimized timing, quantum information can be reliably transferred.

Fig. 9.19 Interconnecting two ion traps using a photonic channel can be achieved by coupling the trapped ions to cavities and transmitting the photon via an optical fiber.

Although the linking protocol and the physics of the interface provide a beautiful method, the cavity technology requires advanced hardware and thus poses severe experimental problems. Therefore, yet a different technique to interconnect two individually stored ions was proposed by I. Cirac and P. Zoller. For this purpose they consider ions trapped in an array of individual traps (cf. Fig. 9.20) and they envision for the coupling and transfer of quantum information another ion stored somewhat above this trap array in such a way that it can be moved like a reading and writing head to be coupled to individual ions. According to this proposal the ion–ion coupling is achieved using a focused laser beam that "pushes" the ions in a state-dependent way and thus achieves entanglement. Moving the head then to a different storage site of the trap array would allow one to create entanglement between ions at two different locations. This technology seems a viable approach, similar to the hard disk drives available with classical computers. However, trapping the "head ion" and moving the entire head is experimentally challenging and has not been addressed yet. Instead, the Boulder group has come up with an approach that makes use of moving ions, however, without moving any head device at all. This requires a somewhat different philosophy: Instead of using ions stored at fixed locations, it was proposed to use the ions as moving memory altogether and thus the trap structure needs to be completely different. For this reason and in order to

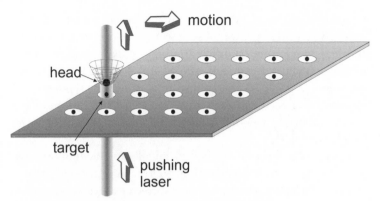

Fig. 9.20 A single ion is considered as a movable head that can be coupled to individual ions of a trap array by the use of a coupling ("pushing") laser.

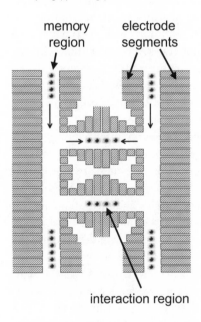

Fig. 9.21 Diagram of the quantum-charged-coupled device (QCCD) proposed by the Boulder group. Ions are stored in the memory section and moved to the interaction region for logic operations. Thin arrows show transport and confinement along the local trap axis (from [12]).

build up a large(r)-scale quantum processor the Boulder group proposed a "quantum-charged-coupled-device" (QCCD) that consists of many interconnected segmented ion traps. The idea is to shuttle ions between different locations and traps to carry the quantum information. This would allow communication between sets of ions [12]. In such a structure one could distinguish between a loading area, a cooling and logic region. It it conceivable that shuttling the ions back and forth between different areas of such segmented trap structure would allow one to interconnect different groups of ions serving as memory and accumulator part similar to what is done on a conventional processor. Currently, several groups are pursuing this approach with the goal to build and operate a small ion chip. With this technique, in fact, scaling the ion-trap quantum information processor seems feasible after all.

A major ingredient for further scaling of any quantum information processor is the implementation of error correction. Whereas in classical information processing error correction is very well known and implemented, it was not clear for some time whether error correction would actually work with quantum processes. The reason for this is simply that quantum information cannot be copied like classical information and therefore, classical error-correction protocols that usually rely on redundant encoding do not work. Fortunately and surprisingly, it has been known now for several years that with the use of entangling operations, quantum error-correction protocols can actually work and the first experiments towards their implementation have already been carried with NMR-based systems and very recently even with trapped ions by the Boulder group.

For a reliable operation of extended quantum information processing it will be indispensable to implement error-correction protocols routinely. It appears that the information of a single two-level system needs to be encoded into five or even seven physical qubits and a certain number of quantum gate operations and measurements will be necessary to keep that information stored and "alive" as a logical qubit. Scaling a quantum information processor will then require the coupling of logical qubits and subsequently gate operations between them. At this time there are worldwide efforts to develop the right computer architecture with many of the systems mentioned above. For trapped ions such ideas are pursued by Isaac Chuang from MIT and the experimental groups are trying hard to implement them.

9.12 The future

... is bright! Quantum information processing has turned out to be a major research area in many areas of physics. Aside from the fact that one might actually be able to build, some time in the future, a device that is capable of number crunching with incredible efficiency, the required amount of quantum engineering and the control of quantum states on a macroscopic scale is a challenging and intriguing task in itself. Speaking of quantum information processing one is often asked the question when one would be able to actually buy a quantum computer, meaning a device that could replace the standard PC. While this is certainly quite a way off, small quantum computers will be available in the not too distant future. For example, for the purpose of long-distance quantum communication, small nodes with only a few qubits would be sufficient as quantum repeaters. For quantum simulations, processors with several tens of qubits would suffice and they seem to be in reach within the next ten years.

While quantum information processors with trapped ions seem bulky and often balky at present, they offer a viable route towards larger devices. On the other hand, with such devices we may be able to get to the age of a quantum "tube computer" and they will provide us with the playground on which we can try out and invent more algorithms, better quantum processing and even do new physics. From a personal point of view, one of the most striking results of quantum information processing is the possibility of error correction, a feature conceived only during the last decade. Using quantum coherences and their possibly uninterrupted, protected or corrected, dynamical evolution will lead to new schemes for precision measurements and offer new ways of bringing the quantum world to a larger scale.

References

1 R. P. Feynman. Simulating physics with computers. *Int. J. Theor. Phys.* **21**, 467 (1982).

2 D. Deutsch. Quantum theory, the Church-Turing principle and the universal quantumcomputer. *Proc. R. Soc. Lond. A* **400**, 97 (1985).

3 P. Shor, in Proceedings, *35th Annual Symposium on Foundations of Computer Science*, IEEE Press, Los Alamitos, CA 1994; P. Shor, SIAM J. Comp. **26**, 1484 (1997); see also A. Ekert and R. Josz[a], Rev. Mod. Phys. **68(3)**, (1996).

4 L. K. Grover. Quantum mechanics helps in searching for a needle in a haystack. *Phys. Rev. Lett.* 79, 325 (1997).

5 D. P. DiVincenzo. Dogma and heresy in quantum computing. *Quantum Information and Computation* 1, Special 1–6 (2001).

6 J. I. Cirac, P. Zoller. Quantum computations with cold trapped ions. *Phys. Rev. Lett.* **74**, 4091 (1995).

7 D. M. Greenberger, M. A. Horne, A. Zeilinger, *Phys. Today* **46**, N° 8, 22 (1993).

8 C. F. Roos, M. Riebe, H. Häffner, W. Hänsel, J. Benhelm, G. P. T. Lancaster, C. Becher, F. Schmidt-Kaler, R. Blatt. Control and Measurement of Three-Qubit Entangled States. *Science* **304**, 1478 (2004).

9 C. Bennett et al., Teleporting an unknown quantum state via dual classical and EPR channels. *Phys. Rev. Lett.* **70**, 1895 (1993).

10 M. Riebe, H. Häffner, C. F. Roos, W. Hänsel, J. Benhelm, G. P. T. Lancaster, T. W. Körber, C. Becher, F. Schmidt-Kaler, D. F. V. James, R. Blatt. Deterministic quantum teleportation with atoms. *Nature* **429**, 734 (2004).

11 M. D. Barrett, J. Chiaverini, T. Schaetz, J. Britton, W. M. Itano, J. D. Jost, E. Knill, C. Langer, D. Leibfried, R. Ozeri, D. J. Wineland. Deterministic quantum teleportation of atomic qubits. *Nature* **429**, 737–739 (2004).

12 D. Kielpinski, C. Monroe, D. J. Wineland. Architecture for a large-scale ion-trap quantum computer. *Nature* **417**, 709 (2002).

10
Quantum theory: a challenge for philosophy!

Michael Esfeld

Quantum theory is a challenge for the philosophy of nature in three aspects. Quantum systems, as a rule, are not localized at points or arbitrarily small regions in space. They are not single systems characterized by attributes independent of each other, they are rather linked with one another through relations of the entanglement of states. And they are not individuals. Quantum theory suggests a philosophy of nature that emphasizes the relations. Bell's theorem together with the formalism of quantum theory and the results of the experiments entitle us to the claim that these relations are not based on intrinsic properties. Quantum theory opens up the epistemological perspective to overcome the dualism of relations, which are accessible to us, and the intrinsic properties behind them, which are unknown to us. A philosophy of nature on the basis of quantum theory can be set out in a way that it is consistent with everyday realism.

10.1 Challenges for philosophy of nature

Quantum theory is a challenge for philosophy on two levels: on the level of the philosophy of nature — that is the level of our understanding of nature — and on the level of epistemology — that is the level that considers the question in which way we can gain knowledge about nature. At first I will go into the philosophy of nature and outline the changes that result due to quantum theory. Then I shall deal with epistemology and show how perspectives for the solution of a classical philosophical problem arise from quantum theory. Finally I will discuss the relationship between the view of nature that is suggested to us by quantum theory and the commonsense world, which is characterized by macroscopic objects.

Albert Einstein was known to be sceptical about quantum theory. His understanding of nature, which motivated his objections against quantum theory, was formulated clearest in an essay from 1948. There, he says:

> "Further, it appears to be essential for this arrangement of the things introduced in physics that, at a specific time, these things claim an existence independent of one another, insofar as these things 'lie in different parts of space'. Without such an assumption of the mutually independent existence (the 'being-thus') of spatially distant things, an assumption which originates in everyday thought, physical thought in the sense familiar to us would not be possible." (Howard (1985), p. 187, Einstein (1948), p. 321)

This independence of the existence was also named the "separation principle" or the "separation hypothesis" by Einstein in a letter to Schrödinger from June 19, 1935 (quoted in Howard (1985), p. 179–180). Today we speak about the *principle of separability*. We assume that a multitude of single things exists on the fundamental level of nature. These are not necessarily particles. They can also be points where field attributes occur. We name these things *physical systems*. The statement that the physical systems are independent of each other has the following meaning: *each one of the systems has its basic characteristic attributes independently of all the other systems*. This means it has these attributes independently of the question whether other physical systems in fact exist or not. This is therefore a matter of an inner property of a system. Such attributes are named *intrinsic properties* in philosophy.

Intrinsic properties can be acquired through causal relations. The point that matters is exclusively the possibility to have such properties independently of the existence of other things. Consider the property of being a grain of sand. There might be a chain of causes that led to the existence of each single grain of sand in our world, which involves an undetermined number of other things. But this cannot prevent the fact that if there were only one single thing in the world, this thing could be a grain of sand. For this reason, being a grain of sand is an intrinsic property.

Beyond this, the principle of separability says the following: *the relations that exist between physical systems are determined by the intrinsic properties of the respective systems.* Let us assume that mass is an intrinsic property. Paul has a mass of 80 kg and Peter has a mass of 70 kg. In this case, the relation that Paul is heavier than Peter is determined by the masses that Paul and Peter have independently of each other. In analogy, one can assume that the spatial distance between two systems is determined by the position of each one of the systems; the position is an intrinsic property in the sense that the position of one system is independent of the positions of all other systems. The physical concept of separability can be summed up in the following way: *each system taken just by itself has a state that completely indicates the values of the time-dependent properties of this system. The state of an entire system that consists of several subsystems is determined by the states of the subsystems.* Only those properties of which the value can change during the existence of the system are time dependent — like for example position and momentum, but not mass and charge. These properties can also be called "state-dependent properties". The state of a system at a certain time is the way in which this system has time-dependent properties at this time.

Beyond the principle of separability, Einstein, in continuation of the quotation above ((1948), p. 321–322), takes on another principle that concerns the change of the states of physical systems: *the principle of local action.* Causal effects (interactions, forces) propagate from one point to a neighboring point with a limited velocity (of which the speed of light is the upper limit). The principle of local action presupposes the principle of separability: it concerns changes in systems that all taken by themselves have a state in the given sense. The principle of local action is a kind of locality condition. Sometimes it is identified with the principle of locality without getting more precise (for example Howard (1985). p. 173, 179). The identification of local action with locality, however, can lead to some misunderstanding in the interpretation of the quantum theory. In this chapter, "interpretation of quantum theory" means to work out a suggestion about the way the world is on the fundamental physical level, provided that quantum theory is correct.[1]

Einstein says in the quotation above that what is known today as the principle of separability originates from everyday thinking. Philosophically we can trace this principle back to Aristotle. A central statement in the metaphysics of Aristotle is that the world consists of a multitude of single things (substances), each of them characterized by intrinsic properties (for example in *Categories*, Chapter 5, and *Metaphysics*, book VII). We can find a clear contemporary formulation of this view of nature in the American philosopher David Lewis (1941–2001):

> We have geometry: a system of external relations of spatio-temporal distance between points. Maybe points of spacetime itself, maybe point-sized bits of matter or aether or fields, maybe both. And at those points we have local qualities: perfectly natural intrinsic properties which need nothing bigger than a point at which to be instantiated. For short: we have an arrangement of qualities. And that is all. ... All else supervenes on that. (Lewis (1986), p. IX–X)

The meaning of supervenience in this context is that everything that exists in our world is fixed by the distribution of intrinsic properties at points.

Quantum theory is a challenge for this philosophical tradition in which Einstein stands, and which spans from Aristotle until today. We can distinguish three aspects of this challenge. At first, some attributes of each quantum system depend on each other in the sense that not all of these attributes can have a definite numerical value at a given time. The best known example of such incompatible or complementary attributes are position and momentum: the closer the value of the position gets towards a definite numerical value, the larger is the uncertainty of the value of the momentum and vice versa. This is the content of Heisenberg's uncertainty relation. No state of a system exists, in which the product of the uncertainty of the position and the momentum can fall below a certain value. Uncertainty means in this context the difference from a definite numerical value. Instead of having a value like a point, we are dealing with a distribution of values.

1) Compare to Section 1.10

This inequality is sometimes referred to as "indeterminacy relation". This expression can be misleading: there is no indeterminacy in the sense of imprecision. The Heisenberg inequality gives a precise lowest limit for the product of the dispersion of position and momentum. The expression "uncertainty relation" can be misleading as well: this inequality has nothing to do with the uncertainty of observers about the position or the momentum of the system. Uncertainty is something that concerns these properties as such, and that can be precisely determined (compare Brown and Redhead (1981)). The most important consequence of this mutual dependence of attributes within a quantum system is the fact that *quantum systems, as a rule, are not localized*. Their attributes — even their state-independent properties like mass and charge — do not exist at points or in arbitrarily small regions in space.

This mutual dependence of attributes within a quantum system has far-reaching consequences. When we measure a state-dependent attribute of a quantum system, the measuring instrument does not register an attribute that the quantum system has independently of its interaction with the measuring instrument. It is rather the case that the quantum system acquires a certain value of the corresponding attribute only relative to the fact that the measuring instrument acquires the attribute of showing a certain value. We thus do not have intrinsic properties but relational attributes, which consist in a correlation between a quantum system and a measuring apparatus.[2]

The attack on intrinsic properties, which is included in quantum theory, goes even further. When we take a look at two or more quantum systems, each of them having at least two attributes that mutually depend on each other in the sense described above, the formalism of quantum theory allows that the *states* of these systems are *entangled with each other*. Let us recall the simplest case of entanglement.[3] The simplest quantum physical attribute is the spin of a quantum system, a kind of a characteristic angular momentum. Electrons, for example, are systems with spin ½. In this case, the spin components in each one of the three spatial directions can take only two definite numerical values, $+1$ and -1. In the following, when the spin is mentioned without any further explanation, this is always about the spin components. We name these values "spin plus" and "spin minus". We

2) See also Section 8.3.
3) See also Section 2.2.

now consider two systems with spin ½, which are emitted together from a source and that are moving away from each other in opposite directions, so that they no longer interact. Nevertheless, neither of the two systems has a spin state independent of the other system. The global state of both systems is a superposition of "first system spin plus and second system spin minus" with "first system spin minus and second system spin plus"; this state is known as a singlet state.[4] One of these two components becomes manifest in a measurement. The result is either "first system spin plus relative to second system spin minus (relative to the corresponding readings of the measuring instruments)" or "first system spin minus relative to second system spin plus (relative to the corresponding readings of the measuring instruments)". This example goes back to Bohm (1951), p. 611–622. Conceptually similar examples can be constructed with the attributes of the position or the momentum of two systems (Einstein, Podolsky and Rosen (1935)). In these cases we are dealing with an infinite number of values instead of two, which get into the entanglement of the states. Following the work of Einstein (1935), the correlations of the entanglement of states are known as Einstein–Podolsky–Rosen correlations.

The entanglement of states means, generally formulated, that none of the systems by itself has a value for the state-dependent attributes in question — neither has it a value with an uncertainty that depends on other attributes of the same system. Quantum theory, instead, gives us correlations between the relevant systems with respect to their state-dependent attributes. These correlations are independent of the spatial or spatial-temporal distance of the systems in question. These correlations are not a case of causality; because the condition for a causal connection between the changes of the states of physical systems is the existence of a state for each one of the systems in the sense of the principle of separability. The correlations of the entanglement of states, on the other hand, break this principle. Moreover, the entanglement of states means a reversal of the principle of separability. Instead of each of the relevant systems having a state that is independent of all the other systems, only the systems taken together can have a well-defined state: only the entire system is in a pure state. From the state of the entire system — and only from this one — it is

---

4) Compare to Section 2.4.

defined whatever is true about the state-dependent attributes of the subsystems, namely that only relations between the subsystems exist in the sense of the Einstein–Podolsky–Rosen correlations we have mentioned.

Starting from correlations between quantum systems and measuring instruments over Einstein–Podolsky–Rosen correlations in between quantum systems, we can extend the discussion of the latter correlations so far that we finally get to a whole network of such correlations. Entanglements of states like the examples mentioned here are from the point of view of the formalism of quantum theory not exceptional cases. Whenever we look at a complex system that has several quantum systems as components, we can expect the states of these quantum systems to be entangled with each other. This is also true of the whole of nature on the level of quantum systems. Thus, in the end all state-dependent attributes in the form of the mentioned correlations are therefore only completely determined from the overall state of all quantum systems together. It is due to this that one speaks about holism in respect of quantum theory: the quantum systems are chained together by relations, which in no way can be put back to something that can be attributed to the single quantum systems independent of each other (see Esfeld (2004)).[5] When starting from classical physics and the principle of separability, one can still hold on to single physical systems in quantum mechanics. Also, experiments can be carried out with single quantum systems. But whatever is considered to be intrinsic properties in classical physics has to be transformed to correlations between these systems.

The following objections can be made against all that was said so far: the entanglements of states only concern the state-dependent or time-dependent attributes. But quantum systems also have state-independent properties like, for example, mass and charge. These properties also belong to the basic attributes of quantum systems. Properties like mass and charge, however, are no basis from which the correlations between the quantum systems could be defined. Whether properties like mass and charge are really intrinsic is, furthermore, questionable. A point charge is embedded in a whole field. In connection with the general relativity theory, one can even argue that the mass is also a relational attribute (compare to Teller

---

5) Compare to Section 2.8.

(1991), section VI and VII). Apart from this, being able to deduce properties like mass and charge within the formalism of quantum theory would be desirable, since quantum theory is our fundamental physical theory. Some strategies indeed exist, which could achieve this. The idea there is to get to state-independent properties like mass and charge on the basis of attributes that are relational in the sense described above.

The fact that quantum systems do not satisfy the principle of separability has an additional consequence, which is a third challenge for philosophy: *quantum systems are not individuals.* Reasons for the missing of individuality in the case of boson-type quantum systems can be given independently of the entanglements of states. The entanglements of states, however, show that all types of quantum systems — despite the distinction between bosons and fermions and the Pauli principle, which holds for the latter — are not individuals. When, in quantum mechanics, we take a look at several quantum systems of the same type, with states that are entangled with each another, these systems are indistinguishable. No attributes exist at all — not even conditional probability distributions of any kind — by which a system is distinct from any other such system (see French and Redhead (1988)). From this it follows that quantum systems have no identity in time. We cannot mark a quantum system and then recognize it again. The systems of quantum mechanics (electrons, protons, neutrons, etc.) are single systems and an entire system always contains a definite number of such systems. In quantum mechanics, the correlations of the entanglement of states always exist in between a definite number of single systems. In the examples mentioned above we always deal with two systems. Furthermore, each of these systems is a subject for the predication of attributes and thus a carrier of attributes — be it attributes like "is entangled with other systems", "has spin plus when the other system has spin minus" and so forth.

Quantum field theory goes even further. What single physical systems are according to quantum mechanics, the same are attributes of quantum fields according to quantum field theory. Quantum field theory formulates for each kind of elementary systems (such as electrons and the likes) a field that stretches over the whole of space-time. Electrons and the like are field quanta that are mathematically represented as a particle number operator. There is no need for a field to be in a state in which it has a definite number of field quanta. It could, for example, be in a state that is a superposition of five and seven electrons. The correlations known from quantum mechanics appear in quantum field theory as correlations between the conditional probability distributions of field operators at space-time points. The quantum field theory confirms the relevance of these correlations. These correlations even exist in the vacuum state, as is emphasized by the Reeh–Schlieder theorem (see Redhead (1995)). The quantum field theory therefore disposes of the single-quantum systems as the carriers of attributes but it holds on to the correlations, which are now formulated with reference to points of the space-time.

The quantum theory is thus, from three points of view a challenge for the philosophy of nature:

- *Locality*: Quantum systems and their attributes, as a rule, are not localized at points or any small regions of the space or the space-time.

- *Separability*: The fundamental time-dependent attributes of quantum systems, like position, momentum and spin in the sense of the spin components, are not intrinsic properties that would belong to every quantum system itself. Regarding attributes, the quantum systems, instead of having intrinsic properties, have first of all relations in the sense of the mentioned correlations. These relations of the entanglement of states are, in the end, completely defined only for the overall state of all quantum systems together.

- *Individuality*: Quantum systems are not individuals. In the quantum field theory, even the single quantum object, as a subject for the predication of attributes, is no longer necessary.

The greatest challenge is the lack of the separability. The lack of the individuality is a consequence of the nonseparability. At this point we should bear it firmly in mind that the quantum theory, instead of suggesting a philosophy of nature of single things with intrinsic properties, proposes a philosophy of nature that emphasizes the relations. Before following up this point any further with respect to the philosophy of nature, I would like to go into the epistemological perspectives that are opened up by the quantum theory.

10.2 Epistemological perspectives

Einstein considers quantum theory as being incomplete. He accepts the predictions, but he holds that even quantum systems satisfy the principle of separability: intrinsic properties of quantum systems exist, which underlie the correlations. Einstein was envisaging a physical theory that keeps the predictions of quantum theory and describes the intrinsic properties that form the basis of the observed effects.

Let us for once assume that such intrinsic properties exist. How should we be able to recognize them? When the physical things are independent of each other and from us in the way demanded by the principle of separability, how could we have any access to them? We gain some knowledge about the physical things in a way that they get into a relation with us or with our measuring instruments. In this way, however, we can only find out something about the relations between physical things but nothing about their inner nature (their intrinsic properties). In this point lies the epistemological problem for the philosophy of nature from Aristotle to Einstein and David Lewis, which is based on the principle of separability. The Australian philosopher Frank Jackson has recently formulated this problem in the following way:

> When physicists tell us about the properties they take to be fundamental, they tell us what these properties *do*. This is no accident. We know about what things are like essentially through the way they impinge on us and our measuring instruments. … [This] leads to the uncomfortable idea that we may know next to nothing about the intrinsic nature of the world. We know only its causal cum relational nature. … I think we should acknowledge as a possible, interesting position one we might call Kantian physicalism. It holds that a large part (possibly all) of the intrinsic nature of our world is irretrievably beyond our reach, but that all the nature we know about supervenes on the (mostly or entirely) causal cum relational nature that the physical sciences tell us about. (Jackson (1998), p. 23–24)

Jackson is not supporting any relativism. He is not claiming that we only know the way in which physical things behave relative to us, or that physics depends on a conceptual scheme, which is only valid in our culture. The problem is that we can only gain knowledge about the things through the relations these things get into with each other or with us or with our measuring instruments. The relations to us or to our measuring instruments are only a special case of the relations about which physics formulates general laws. But the point is that these laws only concern relations, even though they are relations of things between each other. Therefore, an objection suggests itself: we cannot gain any knowledge about the intrinsic properties of things. It is quite appropriate to refer to Kant in this context, as Jackson does: We cannot know what the physical things are in themselves.

When supporting a philosophy of nature based on the principle of separability, one is confronted with the epistemological problem of the way in which we should be able to gain knowledge of the intrinsic properties of the objects that are independent of each other. Quantum theory fits in here as follows: it claims explicitly to describe only relations between physical things, but not intrinsic properties. When the argument that intrinsic properties are unknowable is accepted, one can accept that quantum theory is complete in the sense that it says everything that can be said about quantum systems.

The position that quantum theory is complete is linked to Einstein's antagonist Bohr and his Copenhagen school. Bohr admits the completeness of quantum theory — but at the price that he allows statements referring to quantum systems only relative to the description of an experimental arrangement including a measurement. Therefore, attributes are only assigned to quantum systems relative to experimental arrangements and measurements. Bohr's point of view leaves open two, in principle, different possibilities of interpretation:[6]

1. One can maintain that Bohr does not deny that there is more about quantum systems than the behavior that shows up in the experiments (for example Folse (1985), Chapters 7 and 8). Bohr's point is merely that we only have access to the way in which quantum systems behave relative to our experimental arrangements and measurements. From this point of view, the disagreement between Einstein and Bohr boils down to the following issue: both agree on the existence of an intrinsic nature

6) See also Chapter 3.

of quantum systems beyond the phenomena that become manifest in the experiments. Bohr claims that we can gain knowledge only about these phenomena. Einstein assumes that it is possible to achieve a physical theory that describes the intrinsic nature of quantum systems.

2. One can maintain that it is Bohr's point to deny that anything in quantum systems exists, which goes beyond the behavior that they show in experiments (for example Faye (1991), Chapter 8). In this case, however, one has to admit that Bohr's arguments are not sufficient to support this conclusion. An interpretation that accepts statements about quantum systems only relative to an experimental arrangement, including a measurement, does not have the strength to exclude the possibility that an intrinsic nature of quantum systems beyond the experimentally accessible phenomena exists. From this perspective, quantum theory may be complete in the sense that it says everything that *can be said* about quantum systems. Any reason for the assertion that quantum theory would be complete in the sense that it describes everything that *exists* about quantum systems is missing. To argue against an intrinsic nature of quantum systems beyond what is manifest in the experiments, we have to go beyond Bohr's interpretation to a kind of a realistic or ontological interpretation of quantum theory.

The theorem of John Bell (1964) is decisive for the progress of the debate on the positions of Einstein and Bohr. Bell starts from Einstein's principle of separability and the local action. He shows that based on these principles, there is an upper limit for correlations of the kind that quantum theory deals with. Quantum theory predicts higher correlations between the conditioned probability distributions of quantum systems than are allowed by Bell's theorem.[7] The predictions of quantum theory are confirmed by experiments - in particular experiments of the type made by Aspect et al. (1982), in which any causal interaction between the correlated quantum systems with a velocity that is not faster than the speed of light is excluded.

Bell's theorem can be interpreted in a way that it shows the following: *if quantum systems have an intrinsic nature underlying the*

7) Compare to Section 2.5.

correlations — as expressed by the principle of separability — *there is a certain upper limit for the correlations between quantum systems, which can become manifest in the experiments.* This statement, taken just as it is, is not all that interesting yet. But it becomes interesting due to the modus tollens that it allows: *if the correlations that become manifest in the experiments cross the upper limit that is set by Bell's theorem, the quantum systems have no intrinsic nature underlying the correlations* — as expressed by the principle of separability.

When one accepts the predictions of quantum theory and their experimental confirmation, Bell's theorem, in accordance with this interpretation, speaks against a philosophy of nature of single things that are independent of each other: no intrinsic nature of quantum systems beyond the correlations that are manifest in the experiments exist. But Bell's theorem does not decide the argument between Einstein and Bohr simply in favor of Bohr. It says more than what Bohr's position would allow us to say. Bell's theorem gives reason for the claim that quantum theory is complete in the sense that any intrinsic nature of quantum systems beyond the correlations is excluded. When we interpret Bohr in the sense of (2), Bell's theorem gives us exactly the reason for this philosophical conclusion, which is missing in Bohr. Quantum theory is complete — not only in the sense that it states everything that can be said about quantum systems, but also in the sense that it describes everything that exists about quantum systems.

However, we do not have to conceive Bell's theorem in the presented way. As always in philosophy, this is a matter of plausibility considerations. At first, one can dispute the connection between the principles of probability theory, on which Bell's proof of his theorem is based, and the principle of separability. There are a few voices according to which Bell's theorem concerns only a mathematical aspect that, as such, is not of any philosophic interest (see, in particular, Fine (1982a) and (1982b)). This view, however, can be rejected with good formal arguments (see Müller and Placek (2001)).

Secondly, two objections can be made, which are related to the formalism of quantum theory. It is not necessary for the states of quantum systems to be entangled with each other. There are some cases where we look at two or more quantum systems, of which the states are not entangled with each other. These are cases of product states. There, one can argue that product states fulfil the principle of separability and that they refer to intrinsic properties, at least in some cases. It was already said above, however, that the entanglements of states are generic according to the formalism of quantum theory: whenever we look at a complex system that consists of two or more single quantum systems, we have to expect first of all that the states of these systems are entangled with each other. Thus, basing oneself on the formalism of quantum theory, the situations that have to be explained are not the entanglements of states but the cases of their absence. When, in accordance with quantum theory, intrinsic properties have to be introduced in such a manner that they are deduced in some way from the relations of the entanglement of states, no objection against the stated argument follows. The point of this argument is the statement that no intrinsic properties exist on which the relations of the entanglement of states could be based.

Even in the case of the entanglement of states of quantum systems it is possible to give a description for each of the relevant quantum systems separately. This is a description in terms of what is known as a mixed state (improper mixture) (Espagnat (1971), Section 6.3). This description contains all the information that a local observer could gain separately about each of the quantum systems in question, but it ignores the correlations in which the entanglements of states consist. Therefore, this description does not take all the factors relevant for the predictions of measurement results into account. The description in terms of whatever is regarded as a mixed state is an incomplete description of quantum systems. This is not a case of a description of intrinsic properties a quantum system has independently of the existence of other quantum systems. The availability of such a description therefore is no objection against the considered argument.

Apart from these objections coming from the formalism of quantum theory, one can, thirdly, dispute that the experiments really show a breaking of Bell's theorem. As is the case in any experiment, shortcomings could also be found in the experiments following Bell's theorem. The most important shortcoming is the fact that the detectors are inefficient in all experiments carried out so far. Most of them are correlation experiments with pairs of photons. A remarkable number of photons are not registered by the detectors. It is logically possible to develop models with hidden variables that reproduce the results of the experiments carried out so far without assuming Bell's theorem is violated (see Fine (1982c)). However, one could object that this comes to some kind of a conspiracy, when all hidden variables that agree with separability and local action would influence the detection of the photons just in such a way that the predictions of quantum theory are fulfilled (compare, for example, Maudlin (1994), p. 175–186).

More important than such experimental shortcomings is, fourthly, the following comment: It is logically possible to accept that the experiments show the violation of Bell's theorem, but nevertheless hold on to the idea that quantum systems have an intrinsic nature and thus fulfil the principle of separability. Hidden variables could, for example, exist that consist in interactions faster than the speed of light between quantum systems that are independent of each other (see for example Chang and Cartwright (1993), p. 181–189). Another possibility would be to admit hidden variables with a causality that is directed backwards in time (see, for example, Price (1996), Chapter 9).

The debate about the interpretation of quantum theory confirms the standpoint of the American philosopher Willard Van Orman Quine expressed in his famous essay "Two Dogmas of Empiricism": "Any statement can be held true come what may, if we make drastic enough adjustments elsewhere in the system." (in Quine (1980), p. 43). It is possible to hold on to the idea that quantum systems have an intrinsic nature and thus fulfil the principle of separability, if we are willing to support consequences like interactions faster than the speed of light or a causality that is directed backwards in time. The way to take up Bell's theorem that is suggested here is based on the formalism of quantum theory, the experimental results and the evaluation of the price one would have to pay for an alternative theory to quantum theory, which would try to follow the ontology of classical physics.

The situation is sometimes described in the way that we either have to give up the assumption of locality or the assumption of reality, as a consequence of Bell's theorem and the accompanying experiments (for example Davis and Brown (1986), p. 38–39). Giving up the assumption of locality refers to the possibility to back an alternative to quantum theory — in particular an alternative that allows interactions to be faster than the speed of light. When such an alternative appears to be barely attractive to us, we are told that we will have to give up the assumption of reality. But what could that mean? The lack of clarity of statements, like the one that the assumption of reality has to be given up when taking quantum theory seriously, essentially contributes to the reservations with which quantum theory is received in wide circles outside physics and the philosophy of science to this day. The assumption that reality must be given up refers to Bohr's interpretation. However, we have seen that Bell's theorem does not simply confirm this interpretation. When we accept that Bell's theorem together with the formalism of quantum theory and the results of the experiments show that quantum systems do not fulfil the principle of separability, we have to give up a certain view of reality and replace it with a different one. We have to give up the view of reality according to which physical things are marked by intrinsic properties and therefore fulfil the principle of separability. This position should be replaced with a view of reality that refers to relations that are not based on intrinsic properties.

Some philosophers of science — notably Abner Shimony — go as far as speaking about "experimental metaphysics" in this context (Shimony (1989), p. 27): for some metaphysical problems like, for example, the question whether an intrinsic nature of physical systems exists, experimental results become relevant subsequent to Bell's theorem — although metaphysical problems cannot be decided by experiments. Bell's theorem shows a tight dovetailing of physics, epistemology and philosophy of nature. The challenge for epistemology that comes from Bell's theorem opens up a new perspective at the same time. Here, we have the chance to overcome the gap between a philosophy of nature, which, with the principle of separability, is about things that are independent of each other, and an epistemology that tells us that we cannot have any access to the things as they are independent of each other anyway. This gap can be overcome by a philosophy of nature that deals with the relations between things, because the way in which we get access to these relations is not an epistemological problem. The decisive point is the argument for the idea that no intrinsic nature of things beyond their relations exists; this argument is given to us with Bell's theorem.

10.3 Philosophical chances

The last statement takes us to chances for philosophy, which arise from quantum theory. We should not understand the challenges from quantum theory for the philosophy in a way that long-established philosophical principles — like the locality, the separability and the individuality of things — need to be defended, but as a chance to further develop philosophy in a dialog with physics. This concerns first and foremost a philosophy of nature in terms of relations instead of things that are independent of each other. An ontology of relations is known in occidental philosophy. For example Spinoza (1632–1677), in the first book of the *Ethics*, supports the idea that there is only one substance and that all things are modi of this substance, which means relations within this substance. A philosophy of nature in terms of relations can therefore hold on to the concept of substance: the system of relations as a whole can be a substance. The English philosopher Francis Herbert Bradley (1846–1924) elaborated a philosophy of internal relations. This refers to the position that the basic characteristic attributes of things are relations with each other instead of being intrinsic properties (see, in particular, Chapter 13 in Bradley (1920)). These references are not meant to suggest that quantum theory vindicates the positions of Bradley or Spinoza. These references should only draw the attention of the reader to the fact that the type of philosophical position, which can be maintained on the basis of quantum theory, is already present in the tradition of occidental metaphysics. Today, the question if a philosophy of single things independent of each other or a philosophy of relations is plausible, is no longer a matter of a purely philosophical argumentation, but the results from physics enter into the game. From quantum theory, we can gain a weighty argument in favor of the classical minority position, the philosophy of relations.

A philosophy of relations is in a minority position, because one assumes that relations need a nonrelational basis by which they are determined. This opinion stands also in the background of the quotation from Jackson in the previous paragraph. Jackson (1998) rejects the view that "the nature of everything is relational cum causal, which makes a mystery of what it is that stands *in* the causal relations" (p. 24). The widespread idea expressed here is that causal relations require intrinsic properties, and based on them things are causal agents and exert causal forces. However, this idea cannot be generalized. The idea that relations, in general, need a nonrelational basis is not true. We take a look at spatial-temporal relations between points. When points of space-time exist, one can support the idea that all qualitative attributes of a point exist as relations to other points. Arguing that a point must have intrinsic properties based on which it gets into spatial-temporal relations with other points makes no sense.

Let us now imagine a world in which all physical attributes can be reduced to geometric attributes of space-time points (following the concept that the causal force of gravity can be reduced to space-time attributes, which can be justified based on the theory of general relativity). In such a world, all types of physical attributes can in principle be known. They all consist of relations. No intrinsic properties are necessary, because the relational links are space-time points. It is possible to maintain the idea that Spinoza had such a philosophy of nature in mind based on the physics of his time (Bennett (1984), Chapter 4). Subsequent to the theory of general relativity, Wheeler (1962) set his sights on an actual physical undertaking of this philosophy of nature with a geometrodynamics that recognizes no physical systems in addition to space-time. This program, however, has failed. Nevertheless, we have there a model for a philosophy of nature that is suitable for today, namely a philosophy of nature in terms of relations: the quantum correlations are another example of relations, which not only require no intrinsic properties as a basis, but also exclude such a basis, as can be argued based on Bell's theorem.

Still, one way out remains open for the supporter of intrinsic properties. It is possible to say that an intrinsic nature of quantum systems exists, but that this is independent of the attributes considered by quantum theory and that are relational in the mentioned sense. In other words: intrinsic properties unknown to us exist; but these do not determine the relations described by quantum theory in any way (compare the position ascribed to Kant by Langton (1998)). Intrinsic properties in this sense, by definition cut off from anything that can become manifest, are not affected by an argument like the one discussed in connection with Bell's theorem. Going this way out, however, no reason at all remains as to why one should assume such intrinsic properties. The argument that relations would, in general, require some kind of intrinsic properties of the relational links, as said before, is not sound.

Some authors claim that quantum theory describes a world of correlations without referring to underlying single things as correlates (Mermin (1998); see also Rovelli (1996)). This standpoint appears to be paradoxical. Some good sense can nevertheless be made from this against the background of the philosophical tradition of an ontology of relations. When correlations exist, so do, of course, correlates. But as far as it concerns quantum theory, no correlates exist, which are marked independently of the correlations by intrinsic properties and therefore could be identified independently of the correlations. Concerning quantum mechanics, we can consider systems that quantum mechanics deals with — such as electrons, protons, neutrons, photons and the like —, as being single physical systems. Quantum systems of the same type, however, are indistinguishable — which was already mentioned in Section 10.1. As long as we restrict ourselves to quantum mechanics, we have correlates that can be regarded as being single systems, but not individuals. Concerning quantum field theory, it was also briefly mentioned in Section 10.1 that we can look at points of fields in the space-time as being the links, and between these points we have the correlations.

It has already been claimed quite often that no intrinsic nature of things exists beyond relations that are manifest to us. Such a claim, however, very often ends in phenomenalism or idealism and with this in a gap between epistemology and ontology. This is due to the fact that as a rule, no argument is available that such an intrinsic nature could not exist. Such an argument should come from ontology. Bell's theorem opens the way for such an argument (provided that quantum theory is accepted) by supporting the ontological claim that no intrinsic properties of quantum systems exist that underlie the manifest correlations (in the sense that these correlations would be determined by any kind of intrinsic properties).

We can explain this claim in a way that Sartre speaks of a *phenomenological ontology* (Sartre (1943), Introduction). According to Sartre, there is no difference between the phenomena that become manifest to us in experience, and the world or the reality itself, because no reality characterized by intrinsic properties exists beyond phenomena. The phenomena consist in relations — epistemologically first and foremost in relations between physical things on the one hand and us or our measuring instruments on the other hand. For Sartre, the phenomena are the reality itself in the sense that the reality itself is constituted by relations of the type that are manifest as phenomena. This claim does not imply that the whole reality is manifest to us. This is a *phenomenological* ontology because the phenomena do not stand for anything beyond them; the elements of the ontology are of the phenomena type, namely relations. This is *ontology*, because once this position is elaborated, the relations to us or to our experimental arrangements have no special status within this position: they are like all the other relations. Starting from the correlations that are manifest in the experiments, we get to a network of correlations inside of which correlations between quantum systems and measuring instruments have no special ontological status. Thus, as soon as we have the right to make the negative ontological claim that no intrinsic nature exists beyond the manifest relations, we can also formulate an ontology in positive terms, namely an ontology of relations.

We can summarize the considerations made so far in three steps:

1. What is manifest to us are relations. No intrinsic nature of things exists beyond these relations — in the sense of intrinsic properties that would fix the relations.

2. We therefore know not only the way in which the things behave towards us, but also their true nature.

3. The nature of physical systems consists in the relations in which they stand.

Since Bell's theorem justifies the claim that no intrinsic properties exist beyond the correlations that are manifest in the experiments, Bell's theorem opens up the way for a particular realistic or ontological interpretation of quantum theory.

10.4 Scientific realism and everyday realism

What is the relationship between this philosophy of nature and our everyday understanding of the world? Some physicists and philosophers of science go as far as to claim that quantum theory, together with the superposition principle and the Schrödinger dynamics, is a universal physical theory, which is valid for all physical systems including the everyday objects familiar to us. In accordance with this position, all correlations that are possible due to quantum theory are indeed real, and these correlations are even extended to the everyday objects familiar to us. We take a look at the case of the measurement of the spin in a given direction, where only the results spin plus and spin minus exist. According to quantum theory, the correlations "quantum system spin plus relative to instrument reading spin plus" and "quantum system spin minus relative to instrument reading spin minus" are possible. This position claims that both of these superimposing correlations are in fact real. Consequently no state reduction occurs in the course of measurement in any way. The only thing that happens in a measurement is that the state of the quantum system gets entangled with the state of the measuring system. Referring to the simplest example of an entanglement of states between quantum systems, this means: the singlet state is not going to be reduced in a measurement to either "first system spin plus relative to second system spin minus" or "first system spin minus relative to second system spin plus", but both of these correlations exist further and they extend to the measuring instruments. We are just not realizing these superimposed correlations due to the decoherence[8] (on decoherence, see Chapter 5). In this position one has to admit that none of our other scientific theories including our everyday understanding gives any complete description of the objects in question, because they abstract from most correlations. This position goes back to the work of Everett (1957) in so far as this work considers the Schrödinger dynamics to be the only dynamics of quantum systems, and introduces the concept of a relative state. A contemporary formulation is, for example, Lockwood (1989), Chapters 12 and 13.

In my opinion, this position is too enthusiastic about the novelty that is in quantum theory. With this, a philosophical chance is lost, because now a gap between the epistemology and the philosophy of nature or the ontology occurs again — even though it is different from the one we were discussing so far. When everything consists in correlations of entanglements of states in the way they are described by quantum theory, no definite macroscopic objects with definite macroscopic properties exist. In an epistemological reflection, however, one can support the idea that the reference to things with definite properties, and with that real classical surroundings are necessary, so that we are able to form terms with a definite meaning and then formulate physical theories (for this, see Esfeld (1999)).

On this basis, one can argue for the position that the correlations quantum mechanics deals with are more or less limited to the micro-

---

8) See also Chapter 8

physical domain. A possibility that is then still open is the interpretation of these correlations in the sense of dispositions. This means for the simplest case of entanglements of states, the singlet state: each of the two systems has the disposition to acquire a definite numerical value of the spin in a given direction relative to the other system, which acquires the opposite value of the spin in the same direction. These dispositions have a basis that is not composed of dispositions. Their categorical basis is the overall state of both systems together, which is a pure state. One of these dispositions is actualized in a measurement at the expense of the other. The result is either "first system spin plus relative to second system spin minus" or the result is "first system spin minus relative to second system spin plus" (see for example Shimony (1993), Chapter 11). In which way such a disposition is actualised in a measurement at the expense of the other one and with this, the way in which the entanglement of states is reduced, is an open physical question. Already in Section 10.1 we have mentioned that in a measurement no intrinsic property of a quantum system is registered under any circumstances; it is rather the case that a correlation between an attribute of a quantum system and an attribute of a measuring instrument is created.

In accordance with this position, two kinds of correlations exist: correlations of the entanglement of states, meaning that none of the systems in question is in a pure state, and correlations in such a way that a quantum system acquires a definite numerical value of an attribute relative to the situation that a measuring instrument acquires the attribute to indicate such a definite numerical value; in this case, the quantum system is in a pure state. The experiments following Bell's theorem gave evidence of both kinds of correlations; the correlations between measurement results with a spatial separation are the experimental proof for the entanglement of states.

A position that accepts the reduction of states makes it possible for us to reconcile a philosophy of nature in terms of relations following quantum theory with the everyday realism. We can then say that of course macroscopic things with definite physical properties exist — this is self-evident because they exist independently of the question as to which physical theory about our world will finally turn out to be correct. It will only become philosophically problematic when one tries to make the transition from the everyday realism to some kind of ontology of single things that are independent of each other as the fundamental being, which are consequently characterized by intrinsic properties. In short, it is not the everyday realism, but the philosophy that was connected to it from Aristotle to David Lewis, which we should call into question. Quantum theory makes it appear plausible that this philosophical position is not correct but with this, it does not affect the everyday realism.

It rather appears to be the case that finally the everyday objects are realized as certain configurations of quantum systems — or perhaps certain structures in quantum fields. They could also be realized as arrangements of classical atoms that are single entities independent of each other in the sense of the principle of separability. The way in which these realizations can be precisely understood is a philosophical problem. However, the problem of the way in which the scientific view of nature can be brought together with our everyday understanding is present for just any physical theory. This problem should not lead us to the mistake either to drop the everyday realism and thus to overemphasize the scientific view of nature, or to favor the everyday realism so much that only an instrumental value is granted to scientific theories and the philosophical project to ask for the fundamentals of nature is abandoned.

Regarding the challenge that quantum theory poses for philosophy, it is not relevant how far the entanglements of states really reach. The core of the challenge is this: when we start out from quantum theory as being the fundamental physical theory, the situations that need to be explained are not the entanglements of states but when they are absent — no matter how few or how numerous these cases might be *de facto*, or by how little or how much the entanglements of states might be negligible for all practical purposes. Quantum theory blocks the possibility to start from a multitude of single things independent of each other in the philosophy of nature. The challenge from quantum theory towards philosophy is to develop a philosophy of nature of relations, which are finally only determined by nature as a whole on the level of the quantum systems. The chance for such a philosophy of nature is to avoid the traditional gap between epistemology and ontology.

References

- Aspect, A., Dalibard, J., Roger, G. (1982): "Experimental Test of Bell's Inequalities Using Time-Varying Analyzers", *Phys. Rev. Lett.* **49**, 1804–1807.
- Bell, J. S. (1964): "On the Einstein-Podolsky-Rosen-Paradox", *Physics* **1**, 195–200.
- Bennett, J. (1984): *A Study of Spinoza's "Ethics"*, Hackett, Indianapolis.
- Bohm, D. (1951): *Quantum Theory*, Prentice-Hall, Englewood Cliffs.
- Bradley, F. H. (1920): *Appearance and Reality*, Allen & Unwin, London, 1st edn., 1893.
- Brown, H. R., Redhead, M. L. G. (1981): "A Critique of the Disturbance Theory of Indeterminacy in Quantum Mechanics", *Found. Phys.* **11**, 1–20.
- Chang, H., Cartwright, N. (1993): "Causality and Realism in the EPR Experiment", *Erkenntnis* **38**, 169–190.
- Davies, P. C. W., Brown, J. R. (1986): *The Ghost in the Atom*, Cambridge University Press, Cambridge.
- Einstein, Albert (1948): „Quanten-Mechanik und Wirklichkeit", *Dialectica* **2**, 320–324.
- Einstein, A., Podolsky, B., and Rosen, N. (1935): "Can Quantum-Mechanical Description of Physical Reality be Considered Complete?", *Phys. Rev.* **47**, 777–780.
- Esfeld, M. (1999): "Quantum Holism and the Philosophy of Mind", *J. Consciousness Stud.* **6**, 23–38.
- Esfeld, M. (2004): "Quantum Entanglement and a Metaphysics of Relations", *Stud. Hist. Philos. Mod. Phys.* **35**, 625–641.
- Espagnat, B. de (1971): *Conceptual Foundations of Quantum Mechanics*, Benjamin, Menlo Park.
- Everett, H. (1957): "'Relative State' Formulation of Quantum Mechanics", *Rev. Mod. Phys.* **29**, 454–462. Reprinted in B. S. DeWitt and N. Graham (eds.), *The Many-Worlds Interpretation of Quantum Mechanics*, Princeton University Press, Princeton 1973, 141–149.
- Faye, J. (1991): *Niels Bohr: His Heritage and Legacy. An Anti-Realist View of Quantum Mechanics*, Kluwer, Dordrecht.

- Fine, A. (1982a): "Hidden Variables, Joint Probability, and the Bell Inequalities", *Phys. Rev. Lett.* **48**, 291–295.
- Fine, A. (1982b): "Joint Distributions, Quantum Correlations, and Commuting Observables", *J. Math. Phys.* **23**, 1306–1310.
- Fine, A. (1982c): "Some Local Models for Correlation Experiments", *Synthese* **50**, 279–294.
- Folse, H. J. (1985): *The Philosophy of Niels Bohr*, North-Holland, Amsterdam.
- French, S., Redhead, M. L. G. (1988): "Quantum Physics and the Identity of Indiscernibles", *Br. J. Philos. Sci.* **39**, 233–246.
- Howard, D. (1985): "Einstein on Locality and Separability", *Stud. Hist. Philos. Sci.* **16**, 171–201.
- Jackson, F. (1998): *From Metaphysics to Ethics. A Defence of Conceptual Analysis.* Oxford University Press, Oxford.
- Langton, R. (1998): *Kantian Humility. Our Ignorance of Things in Themselves*, Oxford University Press, Oxford.
- Lewis, D. (1986): *Philosophical Papers*, Vol. 2, Oxford University Press, Oxford.
- Lockwood, M. (1989): *Mind, Brain and the Quantum. The Compound 'I'*, Blackwell, Oxford.
- Maudlin, T. (1994): *Quantum Non-Locality and Relativity*, Blackwell, Oxford.
- Mermin, N. D. (1998): "What is Quantum Mechanics Trying to Tell Us?", *Am. J. Phys.* **66**, 753–767.
- Müller, T., Placek, T. (2001): "Against a Minimalist Reading of Bell's Theorem: Lessons from Fine", *Synthese* **128** (2001), 343–379.
- Price, H. (1996): *Time's Arrow and Archimedes' Point. New Directions for the Physics of Time*, Oxford University Press, Oxford.
- Quine, W. van O. (1980): *From a Logical Point of View*. Harvard University Press, Cambridge (Massachusetts), 2nd edn., 1st edn. 1953.
- Redhead, M. L. G. (1995): "More Ado about Nothing", *Found. Phys.* **25**, 123–137.
- Rovelli, C. (1996): "Relational Quantum Mechanics", *Int. J. Theoret. Phys.* **35**, 1637–1678.
- Sartre, J.-P. (1943): *L'être et le néant. Essai d'ontologie phénoménologique*, Gallimard, Paris.
- Shimony, A. (1989): "Search for a World View which can Accomodate our Knowledge of Microphysics". In: J. T. Cushing and

E. McMullin (eds.), *Philosophical Consequences of Quantum Theory. Reflections on Bell's Theorem*, University of Notre Dame Press, Notre Dame, p. 25–37.

– Shimony, A. (1993): *Search for a Naturalistic World View. Volume 2: Natural Science and Metaphysics*. Cambridge University Press, Cambridge.

– Teller, Paul (1991): "Substance, Relations, and Arguments about the Nature of Space-Time", *Philos. Rev.* **100**, 363–397.

– Wheeler, J. A. (1962): *Geometrodynamics*. Academic Press, New York.

Index